S0-ATI-542

Differential Dynamic Programming

Series

Modern Analytic *and* Computational Methods *in* Science *and* Mathematics

A GROUP OF MONOGRAPHS AND ADVANCED TEXTBOOKS

Richard Bellman, EDITOR
University of Southern California

Published

1. R. E. Bellman, Harriet H. Kagiwada, R. E. Kalaba, and Marcia C. Prestrud, Invariant Imbedding and Radiative Transfer in Slabs of Finite Thickness, 1963

2. R. E. Bellman, Harriet H. Kagiwada, and Marcia C. Prestrud, Invariant Imbedding and Time-Dependent Transport Processes, 1964

3. R. E. Bellman and R. E. Kalaba, Quasilinearization and Nonlinear Boundary-Value Problems, 1965

4. R. E. Bellman, R. E. Kalaba, and Jo Ann Lockett, Numerical Inversion of the Laplace Transform: Applications to Biology, Economics, Engineering, and Physics, 1966

5. S. G. Mikhlin and K. L. Smolitskiy, Approximate Methods for Solution of Differential and Integral Equations, 1967

6. R. N. Adams and E. D. Denman, Wave Propagation and Turbulent Media, 1966

7. R. L. Stratonovich, Conditional Markov Processes and Their Application to the Theory of Optimal Control, 1968

8. A. G. Ivakhnenko and V. G. Lapa, Cybernetics and Forecasting Techniques, 1967

9. G. A. Chebotarev, Analytical and Numerical Methods of Celestial Mechanics, 1967

10. S. F. Feshchenko, N. I. Shkil', and L. D. Nikolenko, Asymptopic Methods in the Theory of Linear Differential Equations, 1967

11. A. G. Butkovskiy, Distributed Control Systems, 1969

12. R. E. Larson, State Increment Dynamic Programming, 1968

13. J. Kowalik and M. R. Osborne, Methods for Unconstrained Optimization Problems, 1968

14. S. J. Yakowitz, Mathematics of Adaptive Control Processes, 1969

15. S. K. Srinivasan, Stochastic Theory and Cascade Processes, 1969

16. D. U. von Rosenberg, Methods for the Numerical Solution of Partial Differential Equations, 1969

17. R. B. Banerji, Theory of Problem Solving: An Approach to Artificial Intelligence, 1969

18. R. Lattès and J.-L. Lions, The Method of Quasi-Reversibility: Applications to Partial Differential Equations. Translated from the French edition and edited by Richard Bellman, 1969

19. D. G. B. Edelen, Nonlocal Variations and Local Invariance of Fields, 1969

20. J. R. Radbill and G. A. McCue, Quasilinearization and Nonlinear Problems in Fluid and Orbital Mechanics

24. D. H. Jacobson and D. Q. Mayne, Differential Dynamic Programming

In Preparation

21. W. Squire, Integration for Engineers and Scientists

22. T. Parthasarathy and T. E. S. Raghavan, Some Topics in Two-Person Games

23. T. Hacker, Flight Stability and Control

25. H. Mine and S. Osaki, Markovian Decision Processes

26. W. Sierpiński, 250 Problems in Elementary Number Theory

27. E. D. Denman, Coupled Modes in Plasmas, Elastic Media, and Parametric Amplifiers

28. F. A. Northover, Applied Diffraction Theory

29. G. A. Phillipson, Identification of Distributed Systems

30. D. H. Moore, Heaviside Operational Calculus: An Elementary Foundation

31. S. M. Roberts and J. S. Shipman, Two-Point Boundary Value Problems: Shooting Methods

Differential Dynamic Programming

David H. Jacobson
Harvard University
Cambridge, Massachusetts

David Q. Mayne
Imperial College of Science and Technology
London, England

American Elsevier
Publishing Company, Inc.

NEW YORK · 1970

AMERICAN ELSEVIER PUBLISHING COMPANY, INC.
52 Vanderbilt Avenue, New York, N.Y. 10017

ELSEVIER PUBLISHING COMPANY, LTD.
Barking, Essex, England

ELSEVIER PUBLISHING COMPANY
335 Jan Van Galenstraat, P.O. Box 211
Amsterdam, The Netherlands

Standard Book Number 444-00070-4

Library of Congress Card Number 72-100398

Printed in the United States of America

Contents

1. Introduction

2. New Algorithms for the Solution of a Class of Continuous-Time Control Problems

3. New Algorithms for the Solution of Bang-Bang Control Problems

4. Discrete-Time Systems

5. Stochastic Systems with Discrete-Valued Disturbances

6. Stochastic Systems with Continuous Disturbances

Preface

It is the authors' intention that this book accomplish the following objectives:

1. To illustrate the use of the principle of optimality, locally, in the neighborhood of a nominal trajectory, in the development of new successive approximation algorithms for determining optimal trajectories for a wide variety of dynamic optimization problems,
2. To provide complete details of the derivations of the algorithms, giving rules for their implementation, and including illustrative and computed examples,
3. To indicate the possible advantages over conventional procedures that these algorithms may possess,
4. To report some generalizations of the deterministic algorithms to problems of optimal stochastic control.

The book is intended for applied scientists whose projects might involve dynamic optimization either in design or implementation, graduate students in engineering and applied physics, and practicing engineers.

The mathematical treatment of the material is not intended to be rigorous; that is, results are not stated in theorem and lemma form. Rather, the reader is encouraged by the use of intuitive reasoning to take part in the derivations of the algorithms and in the interpretation of the theoretical and computational results.

Derivations of necessary conditions of optimality via dynamic programming have been reported by S. E. Dreyfus ("Dynamic programming and the calculus of variations," Academic Press, New York, 1965), who has also indicated the usefulness of the intuitive nature of dynamic programming in stochastic situations. Differential dynamic programming is, however, concerned with the development of *actual numerical techniques* based on the conceptual framework of dynamic programming; in this sense the present volume appears to be unique.

The organization of this book is indicated by the following outline: In Chapter 1 the class of dynamic optimization problems to be considered, is defined. Some of the most important literature references in optimal control theory and related computational procedures are given. The notion of differential dynamic programming is introduced. Chapter 2 is concerned with the development of new algorithms for the solution of optimal control problems whose optimal controls are continuous functions of time. Bang-bang control problems are treated in Chapter 3. In Chapter 4 discrete-time optimal control problems are studied. Chapters 5 and 6 are concerned with the generalization of the algorithms developed in Chapters 2 and 3 to stochastic optimal control problems. In Chapter 7 the significance of differential dynamic programming is stressed, and extensions of the technique are indicated. Appendix A contains a comparison between a differential dynamic programming algorithm and the well-known successive sweep method.

A note on style is in order. Equations are numbered afresh in each section, and references are collected at the end of each chapter.

The authors are particularly indebted to the following people for many useful discussions, both at Imperial College and at Harvard University: Professor J. H. Westcott, Professor A. E. Bryson, Jr., Professor Y. C. Ho, Professor S. E. Dreyfus, J. Allwright, G. F. Bryant, P. M. Newbold, D. W. Norris, J. S. Riordon, I. H. Rowe, J. L. Speyer, D. G. Stone, K. S. Tait, W. M. Turski, D. R. Vaughan.

D. H. Jacobson dedicates his contribution to the memory of Louis Jacobson. He wishes to thank Lynne, Lisa, and Sven for their assistance and encouragement during the preparation of the material. Also, he thanks M. Milner for inspired teaching, sincere friendship, and constant encouragement.

The manuscript was typed, excellently, by Judy Behn.

D. H. JACOBSON

D. Q. MAYNE

December, 1969

Chapter 1

INTRODUCTION

1.1. OPTIMAL CONTROL PROBLEMS

In this book we shall describe differential dynamic programming techniques for determining optimal control of nonlinear, dynamic systems. We shall consider the dynamic systems to be described by either nonlinear ordinary differential equations, or nonlinear ordinary difference equations.† The precise formulation of these two classes of optimal control problems is given in the following two sections.

1.1.1 Continuous-Time Systems

The continuous-time dynamic systems studied in subsequent chapters are assumed to be described by the following finite set of nonlinear ordinary differential equations:

$$\dot{x} = f(x, u; t); \qquad x(t_o) = x_o \tag{1.1.1}$$

$x(t)$ is an n-dimensional vector function of time describing the state of the dynamic system at any time $t \in [t_o, t_f]$; $u(t)$ is an m-dimensional vector function of time that represents the control variables available for manipulation at time $t \in [t_o, t_f]$.‡

The performance of the system is measured by the performance index or "cost functional":

$$V(x_o; t_o) = \int_{t_0}^{t_f} L(x, u; t)\, dt + F(x(t_f); t_f) \tag{1.1.2}$$

f is an n-dimensional, nonlinear vector function of its arguments, while L and F are scalar, nonlinear functions of their arguments. The final time t_f may be given explicitly or implicitly.

---

† That is, either continuous-time or discrete-time, finite-dimensional dynamic systems.

‡ The symbols x and u are also taken to mean $x(t)$ and $u(t)$, respectively. Where the meaning is clear, x and u are sometimes used to denote the entire time functions $x(t)$ and $u(t)$; $t \in [t_o, t_f]$.

The notation $f(x, u; t)$ means that f is a function of x, u and, maybe explicitly, of time t. At a particular time $t \in [t_o, t_f]$, f is a function of x and u.†
Similar remarks apply to V, L, and F.

The variables x and u may be required to satisfy some or all of the following constraints:

$$g(u; t) \leqslant 0 \tag{1.1.3}$$

$$\phi_1(x, u; t) \leqslant 0 \tag{1.1.4}$$

$$\phi_2(x; t) \leqslant 0 \tag{1.1.5}$$

$$\psi(x(t_f); t_f) = 0 \tag{1.1.6}$$

where g is a p-dimensional, nonlinear vector function of u at time t;
ϕ_1 is a q-dimensional, nonlinear vector function of x and u at time t;
ϕ_2 is an r-dimensional, nonlinear vector function of x at time t; and
ψ is an s-dimensional, nonlinear vector function of x at time t_f.‡

All the functions f, L, F, g, ϕ_1, ϕ_2, and ψ are assumed to be three times continuously differentiable in each argument.

The object of the control problem is to choose $u(t)$; $t \in [t_o, t_f]$ so that the constraints (1.1.3) to (1.1.6) are satisfied and V, given by Equation (1.1.2), is minimized.

1.1.2. Discrete-Time Systems

The discrete-time dynamic systems are assumed to be described by the following set of nonlinear difference equations:§

$$x_{k+1} = f_k(x_k, u_k); \qquad x_1 \quad \text{given} \tag{1.1.7}$$

where x_k and u_k are the state and control variables, respectively, at time k.

The cost function for the problem is

$$V_1(x_1) = \sum_{k=1}^{N-1} L_k(x_k, u_k) + F_N(x_N) \tag{1.1.8}$$

† The semicolon is used to separate t from the other arguments, some of which may be time-invariant parameters; some of the u's may be time-invariant control parameters, say.

‡ $p \leqslant m$, $q \leqslant m+n$, $r \leqslant n$, and $s \leqslant n$.

§ Subscript k denotes the kth time interval.

Constraints analogous to Equations (1.1.3) to (1.1.6) are

$$g_k(u_k) \leqslant 0 \tag{1.1.9}$$

$$\phi_{1k}(x_k, u_k) \leqslant 0 \tag{1.1.10}$$

$$\phi_{2k}(x_k) \leqslant 0 \tag{1.1.11}$$

$$\psi_N(x_N) = 0 \tag{1.1.12}$$

The control problem is to determine the sequence of controls

$$\{u_k : k = 1, ..., N-1\}$$

that minimizes V_1, given by Equation (1.1.8), and satisfies constraints (1.1.9) to (1.1.12).

1.2. OPTIMAL CONTROL THEORY†

Problem formulations similar to that given in Section 1.1.1, for continuous-time systems, have been studied in the classical calculus of variations literature for many years [1-6]. However, the quantity $\dot{x}$ appears as the choice variable, and no specific control variables u are considered.

The modern formulation of the continuous-time control problem has been studied by a great many researchers, probably the most important ones being [7-37].

Pontryagin *et al.* [8] and Halkin [9-11] have obtained necessary conditions of optimality for the case where control and endpoint contraints (1.1.3) and (1.1.6) are present. Pontryagin's approach is different from classical calculus of variations, and his use of strong variations in the x trajectory allows the inclusion of control variable constraints. Halkin uses the seemingly powerful concept of reachable sets to prove similar results.

Control problems with state variable inequality constraints‡ have been studied mainly by Berkovitz and Dreyfus [12], Dreyfus [13, 14], and Chang [15]. More recently, Speyer [16] has extended these results.

Bellman's principle of optimality [17, 18] would appear to be very general in concept. Bellman and Kalaba [19], Berkovitz and Dreyfus [12], and Dreyfus [13, 14, 20] are a few who have used the principle of optimality (dynamic programming) to derive necessary conditions of optimality. Kalman [21, 22] is more strongly influenced by Caratheodory's work [6].

---

† This section describes deterministic control theory; stochastic control theory appears in Chapter 5.

‡ Inequalities (1.1.4) and (1.1.5) included.

It is surprising that mathematicians and engineers have not used dynamic programming more freely. Some [14, 23] argue that the classical approach and reachable set theory are preferable, since they produce mathematically rigorous results without the strong smoothness requirement on the optimal cost V^o, demanded by dynamic programming. This they believe is especially true of bang-bang control and fixed-endpoint problems. In Chapter 3 we shall show that dynamic programming can be used effectively to study and solve these problems; and, moreover, clear geometric interpretation is obtained for the Pontryagin adjoint variables.

The discrete-time problem, Section 1.1.2, has been studied by a number of authors; Halkin's derivation [24] of the discrete maximum principle is rigorous. Dreyfus [13] has worked with a discrete-time formulation that is a special case of Equation (1.1.7).†

Conditions of optimality are useful to the engineer if he can use them to suggest methods for actually determining the optimal control. Dreyfus [13], Bellman [17], Bellman and Dreyfus [18], and Bellman and Kalaba [19] have considered some computational aspects of dynamic programming. Kelley [25] and Bryson and Denham [26] are well-known for their work in applying the gradient or steepest descent method to control problems.

In fact many different computational algorithms have appeared; the most important we shall describe briefly in the next section.

1.3. CONVENTIONAL METHODS FOR DETERMINING OPTIMAL CONTROL

1.3.1. Introduction

For the purpose of the following sections, it is assumed that, in addition to Equations (1.1.1) and (1.1.2), only constraint (1.1.6) is present.

In addition to requiring the optimal control function $u^o(t)$; $t \in [t_o, t_f]$ for $x(t_o) = x_o$, the engineer would often like a function G such that

$$u^o(x; t) = G(x; t) \tag{1.3.1}$$

i.e., an optimal feedback controller that would enable him to calculate u^o at time t for any particular x.

1.3.2. Dynamic Programming

It is well-known that the optimal value function $V^o(x; t)$‡ obeys the following Bellman partial differential equation:

$$-(\partial V^o/\partial t)(x; t) = \min_u [L(x, u; t) + \langle V_x^o(x; t), f(x, u; t)\rangle] \tag{1.3.2}$$

† $x_{k+1} = x_k + f(x_k, u_k; k).\Delta t$

‡ The superscript o on V denotes "optimal values".

This partial differential equation (PDE) for V^o is, in general, unsolvable analytically; its solution would, however, not only supply us with $u^o(t)$; $t \in [t_o, t_f]$ for $x(t_o) = x_o$, but also the desired feedback controller, Equation (1.3.1).

At this point it is important to note that the Bellman equation is derived on the assumption that the optimal cost $V^o(x; t)$ has continuous first and second partial derivatives with respect to x and t [14].

The difficulty of numerical solution of Equation (1.3.2) is, in general, enormous, primarily because of the high dimensionality of the equation.†

Bellman and others [18, 19] have suggested polynomial approximations for $V^o(x; t)$‡ in order to reduce storage requirements. Larson [27] has indicated a method whereby high-speed storage requirements are reduced greatly. However, the computer solution of Equation (1.3.2) remains a formidable task for problems where the dimensionality of the state vector x is greater than, say, 3.

1.3.3. The Canonical Equations and Pontryagin's Principle

$$\text{Define } H(x, u, V_x^o; t) = L(x, u; t) + \langle V_x^o, f(x, u; t)\rangle \tag{1.3.3}$$

Assume that the minimization with respect to u in Equation (1.3.2) has been performed, and that u^o has been found as a function of V_x^o and x:

$$u^o(x, V_x^o; t) = C(x, V_x^o; t) \tag{1.3.4}$$

By straightforward differentiation of Equation (1.3.2) with respect to x, and use of Equations (1.3.3) and (1.3.4), it can be shown [28] that the following pair of ordinary differential equations results:

$$\dot{x} = f(x, C(x, V_x^o; t); t) \tag{1.3.5}$$

$$-\dot{V}_x^o = H_x(x, C(x, V_x^o; t), V_x^o; t) \tag{1.3.6}$$

$$x(t_o) = x_o \tag{1.3.7}$$

$$\psi(x(t_f); t_f) = 0 \tag{1.3.8}$$

---

† In general, V^o is a function of the n-dimensional x vector at time t.

‡ This is similar in spirit to our differential dynamic programming (DDP) approach (Section 1.4), where the cost function is approximated by a power series expansion about a nominal, nonoptimal trajectory. However, in DDP, *attention is restricted to the immediate neighborhood of the nominal trajectory.*

Equations (1.3.5) to (1.3.8) constitute a two-point boundary-value problem whose solution yields $u^o(t)$; $t \in [t_o, t_f]$ for $x(t_o) = x_o$; we have thus lost the feedback information contained in Equation (1.3.2), and must be content with an "open loop" answer. The coupled equations (1.3.5) and (1.3.6) are referred to as the canonical equations.

Equations (1.3.5) and (1.3.6) are derived with the assumption that the required partial derivatives of $V^o(x; t)$ exist, and are continuous throughout the state space. In some problems, typically, bang-bang control problems with endpoint equality constraints, these derivatives are discontinuous in certain regions of the state space; however, Pontryagin's principle [8, 22] remains valid for these cases. The principle is written as

$$\dot{x} = f(x, C(x, \lambda; t); t) \tag{1.3.9}$$

$$-\dot{\lambda} = H_x(x, C(x, \lambda; t), \lambda; t) \tag{1.3.10}$$

and

$$H(x, u, \lambda, t) = L(x, u; t) + \langle \lambda, f(x, u; t) \rangle \tag{1.3.11}$$

Where the partial derivatives of $V^0(x; t)$ are continuous, $\lambda(t)$ can be identified as $V_x{}^o(t)$. In portions of the state space where $V^o(x; t)$ does not have continuous derivatives, $\lambda(t)$ cannot be interpreted as the partial derivative of $V^o(x; t)$ with respect to x. This point has been studied by Dreyfus [14], Kalman [22], Shapiro [28], and Fuller [29, 30]. In Chapter 3 we shall show, however, that dynamic programming can be used even when $V^o(x; t)$ does not possess continuous partial derivatives everywhere in the state space; we shall give a useful interpretation for Pontryagin's λ in these problems.

Levine [38, 39], Breakwell *et al.* [40] and Jazwinski [41] have developed computational procedures for solving the two-point boundary-value problem. Basically, the missing boundary conditions at one end of the interval $[t_o, t_f]$ are guessed; their values are then successively changed, in an organized fashion, until the given boundary conditions at both ends of the interval $[t_o, t_f]$ are satisfied. Neustadt [42, 43], Paiewonsky [44], and Knudsen [45] are some who have presented algorithms for the solution of the "minimum time" problem.

A disadvantage of the above techniques is that they require the integration of the coupled equations (1.3.5) and (1.3.6), or their linearized version, in the same direction in time; this can lead to numerical stability difficulties if the equations are ill-conditioned and/or the time interval $[t_o, t_f]$ is large [43]. Kalman [46] has written an interesting paper on computational difficulties in control problems.

1.3.4. Quasilinearization

In the quasilinearization method, nominal functions $x(t)$, $\lambda(t)$; $t \in [t_o, t_f]$ are chosen, which do not satisfy the differential equations (1.3.9) and (1.3.10) The pair of equations (1.3.9) and (1.3.10) is then linearized about these nominal values and the resulting linear, two-point boundary-value problem is solved to obtain new $x(t)$ and $\lambda(t)$ trajectories that satisfy more closely Equations (1.3.9) and (1.3.10). This linearization process is repeated successively until $x(t)$ and $\lambda(t)$ satisfy the canonical equations to the desired accuracy. This method has been described by Schley and Lee [47].

1.3.5. Successive Approximation to the Optimal Control Function

The primary advantage of the successive approximation method is that the canonical equations are effectively decoupled; this allows both the equations to be integrated in the stable direction of motion if the dynamic system under consideration is stable.

a. Gradient Method

In the gradient method [13, 25, 26], a nominal control $\bar{u}(t)$; $t \in [t_o, t_f]$ is chosen and the dynamic equations (1.1.1) are integrated forward from t_o to t_f to yield a nominal trajectory $\bar{x}(t)$; $t \in [t_o, t_f]$. The nominal cost $\bar{V}(x_o; t_o)$ is calculated using Equation (1.1.2). The equation $-\dot{\lambda} = H_x$ is integrated backward along the nominal trajectory from t_f to t_o. A change in control $\delta u = -\varepsilon H_u$ is made; and, if ε is small enough ($\varepsilon > 0$), it can be shown that an improvement in cost is obtained on the application of

$$u(t) = \bar{u}(t) + \delta u(t); \qquad t \in [t_o, t_f] \tag{1.3.12}$$

The rate of convergence of the method is initially rapid, when the nominal trajectory is far from the optimal; however, in the neighborhood of the optimum, convergence becomes intolerably slow.

New life has been given to gradient techniques by the advent of the conjugate gradient method [48], which exhibits much improved convergence on some control problems, while retaining the relative simplicity of the original gradient algorithm. Allwright [49] has studied these and related methods in depth.

b. Second-Variation Methods†

In 1960 the LQP problem‡ was elegantly solved [21], and this stimulated research into second-variation methods. Merriam [50], Mitter [51],

---

† Referred to as "sweep methods" by McReynolds and Bryson [52].

‡ Problem with linear dynamics and quadratic performance criterion.

McReynolds and Bryson [52], Kelley [53], and Bullock and Franklin [54] have developed algorithms that converge to the optimal solution in one step, when applied to LQP problems. In non-LQP problems, convergence, if it occurs, is rapid.

These methods require the backward integration of certain differential equations in addition to the usual $-\dot{\lambda} = H_x$. Disadvantages of these methods are: (1) A vastly increased programming effort, compared with the gradient method, is required. (2) H_{uu}^{-1}—the inverse matrix of second partial derivatives of the Hamiltonian with respect to u—is required to be positive-definite along nonoptimal trajectories. This requirement is very often violated in non-LQP problems, resulting in the failure of the methods to converge to an optimal solution.† (3) Control constraints, Inequality (1.1.3), cannot be incorporated directly.

c. METHOD OF CONVEX ASCENT

Halkin's method of convex ascent [55], based on reachable set theory, is basically a first-order method, though no linearization of the dynamics and cost function is made with respect to u. Little numerical experience with this method has been reported.

1.4. DIFFERENTIAL DYNAMIC PROGRAMMING

Differential dynamic programming is a successive approximation technique, based on dynamic programming rather than the calculus of variations, for determining optimal control of nonlinear systems. In each iteration the system equations are integrated in forward time using the current nominal control; and accessory equations, which yield the coefficients of a linear or quadratic expansion of the cost function in the neighborhood of the nominal x trajectory, are integrated in reverse time, thus yielding an improved control law. This control law is applied to the system equations, producing a new and improved trajectory. By continued iteration, the procedure produces control functions that successively approximate the optimal control function.

The DDP approach was first introduced into optimal control by Mayne [56], though Bernholtz and Graham [57] had applied similar notions to economic scheduling. Mayne [56] primarily considered discrete-time systems, though he did derive a second-order algorithm for continuous-time systems by allowing the discrete control problem formulation to tend to the continuous-time formulation.

---

† Kelley [53] and Bullock and Franklin [54] circumvent this difficulty by adding a suitably large positive quantity to H_{uu}.

The second-order algorithm described by Mayne has the same disadvantages as the second-variation methods, but it requires the integration of one set of vector differential equations less than the algorithms described by Mitter [51] and McReynolds and Bryson [52]. Jacobson [58, 59] has shown that the second-variation methods are approximations to Mayne's algorithm.† Recently, McReynolds [60] obtained an algorithm equivalent to Mayne's.

Jacobson [58, 59, 61, 62] and Jacobson and Mayne [63] have applied differential dynamic programming directly to continuous-time systems, and have obtained new algorithms that are applicable to a wider class of problems than conventional methods.‡ Dyer and McReynolds [64] have obtained results similar to those described by Jacobson [62] for bang-bang control problems. Gershwin and Jacobson [65] have recently extended the algorithms described by Jacobson [61] to discrete-time problems.

Mayne [66–68] and Westcott *et al.* [69] have applied differential dynamic programming to stochastic control problems.

In the following two sections the basic differential dynamic programming equations are obtained for continuous-time and discrete-time control problems.

1.4.1. Continuous-Time Systems

It is well-known that the optimal cost $V^o(x; t)$ satisfies the following PDE:

$$-(\partial V^o/\partial t)(x; t) = \min_u [L(x, u; t) + \langle V_x^o(x; t), f(x, u; t)\rangle] \quad (1.4.1)$$

Assume that the optimal control $u^o(t)$; $t \in [t_o, t_f]$ is unknown but that a nominal control $\bar{u}(t)$; $t \in [t_o, t_f]$ is available. On application of this nominal control, a nominal state trajectory $\bar{x}(t)$; $t \in [t_o, t_f]$ is produced by Equation (1.1.1). The nominal cost $\bar{V}(x_o; t_o)$ is calculated using Equation (1.1.2).

Equations (1.1.1), (1.1.2), and (1.4.1) may be written in terms of the nominal trajectory by setting

$$x = \bar{x} + \delta x \quad (1.4.2)$$

$$u = \bar{u} + \delta u \quad (1.4.3)$$

where δx and δu are the state and control variables, respectively, measured with respect to the nominal quantities $\bar{x}$ and $\bar{u}$; they are not necessarily small quantities.

---

† Appendix A.

‡ Chapters 2 and 3.

Equations (1.1.1), (1.1.2), and (1.4.1) become

$$(d/dt)(\bar{x}+\delta x) = f(\bar{x}+\delta x, \bar{u}+\delta u; t); \qquad \bar{x}(t_o) + \delta x(t_o) = x_o \quad (1.4.4)$$

$$V(x_o; t_o) = \int_{t_o}^{t_f} L(\bar{x}+\delta x, \bar{u}+\delta u; t)\, dt + F(\bar{x}(t_f) + \delta x(t_f); t_f) \quad (1.4.5)$$

$$-\partial V^o/\partial t(\bar{x}+\delta x; t) = \min_{\delta u} [L(\bar{x}+\delta x, \bar{u}+\delta u; t) + \langle V_x{}^o(\bar{x}+\delta x; t), f(\bar{x}+\delta x, \bar{u}+\delta u; t)\rangle] \quad (1.4.6)$$

We assume that the optimal cost is sufficiently well-behaved to allow a power series expansion in δx about $\bar{x}$:

$$V^o(\bar{x}+\delta x; t) = V^o(\bar{x}; t) + \langle V_x{}^o, \delta x\rangle + \tfrac{1}{2}\langle \delta x, V_{xx}^o\, \delta x\rangle \quad \text{plus higher-order terms} \quad (1.4.7)$$

The optimal cost at $\bar{x}; t$ is

$$V^o(\bar{x}; t) = \bar{V}(\bar{x}; t) + a^o(\bar{x}; t) \quad (1.4.8)$$

where a^o is defined as the difference between the optimal cost $V^o(\bar{x}; t)$, obtained by using the optimal controls $u^o(\tau) = \bar{u}(\tau) + \delta u^o(\tau)$; $\tau \in [t, t_f]$, and the nominal cost $\bar{V}(\bar{x}; t)$, obtained using the nominal controls $\bar{u}(\tau)$; $\tau \in [t, t_f]$.

From Equations (1.4.7) and (1.4.8),

$$V^o(\bar{x}+\delta x; t) = \bar{V} + a^o + \langle V_x{}^o, \delta x\rangle + \tfrac{1}{2}\langle \delta x, V_{xx}^o\, \delta x\rangle \quad \text{plus higher-order terms in } \delta x \quad (1.4.9)$$

Parameters on the right-hand side of Equation (1.4.9) are evaluated at $\bar{x}; t$. Equation (1.4.6) (with Equation (1.4.9) substituted in) is, in general, awkward to use because of the possibly infinite storage space required for the parameters of the power series expansion. However, if δx is sufficiently small, then Equation (1.4.6) can be expanded to, say, second-order in δx, with error $0(\delta x^3)$. To ensure that all terms up to second-order are present, it is necessary to consider $V^{\circ}(\bar{x}+\delta x; t)$ and $V_x^{\circ}(\bar{x}+\delta x; t)$ expanded only to second-order in δx, i.e.,

$$V^{\circ}(\bar{x}+\delta x; t) = \bar{V} + a^{\circ} + \langle V_x^{\circ}, \delta x\rangle + \tfrac{1}{2}\langle \delta x, V_{xx}^{\circ}\,\delta x\rangle \qquad (1.4.10)$$

and

$$V_x^{\circ}(\bar{x}+\delta x; t) = V_x^{\circ} + V_{xx}^{\circ}\,\delta x + \tfrac{1}{2}\,V_{xxx}^{\circ}\,\delta x\,\delta x \qquad (1.4.11)^{\dagger}$$

(In works by Jacobson [59, 61, 62] and Jacobson and Mayne [63], the $V_{xxx}^{\circ}\,\delta x\,\delta x$ term is omitted. That this is legitimate is demonstrated in Chapter 2.)

V°, given by Equation (1.4.10), is the optimal cost when starting in state $\bar{x}+\delta x$ at time t if either (1) the nominal trajectory is sufficiently close to the optimal one (if $\bar{u}(\tau)$ is close to $u^{\circ}(\tau)$; $\tau \in [t, t_f]$, then the minimizing δu will be small and, from Equation (1.4.4), the resulting δx will be small); or (2) the problem is LQP.

Using Equations (1.4.10) and (1.4.11), Equation (1.4.6) becomes

$$-\frac{\partial \bar{V}}{\partial t} - \frac{\partial a^{\circ}}{\partial t} - \left\langle \frac{\partial V_x^{\circ}}{\partial t}, \delta x\right\rangle - \tfrac{1}{2}\left\langle \delta x, \frac{\partial V_{xx}^{\circ}}{\partial t}\,\delta x\right\rangle = \min_{\delta u}\,[L(\bar{x}+\delta x, \bar{u}+\delta u; t)$$

$$+ \langle V_x^{\circ} + V_{xx}^{\circ}\,\delta x + \tfrac{1}{2}\,V_{xxx}^{\circ}\,\delta x\,\delta x, f(\bar{x}+\delta x, \bar{u}+\delta u; t)\rangle] \qquad (1.4.12)$$

Equation (1.4.12) is valid on the assumption that the δx, produced by δu acting through Equation (1.4.4), is small. If this is not so, then either of the following techniques is used:

1. The minimizing, new control, $\bar{u}+\delta u$ is applied to Equation (1.4.4) only over a small time interval, so that the resulting δx is small. This procedure is described in Section 2.2.2 and is used in Chapters 2 and 3, as well as in the works by Jacobson [58, 61, 62], Jacobson and Mayne [63], and Gershwin and Jacobson [65].

2. The min operation is performed in a restricted sense, i.e., the full minimizing δu is not calculated; only a small step in u, toward the minimum, is taken. This procedure is used in Appendix A and in the works by Mayne [56] and Jacobson [59].

---

† $V_{xxx}^{\circ}\,\delta x\,\delta x \equiv \sum_{i=1}^{n} \sum_{j=1}^{n} V_{xx_ix_j}^{\circ}\,\delta x_i\,\delta x_j$.

For simplicity, in subsequent chapters, the superscript ° on V will be omitted. Equation (1.4.12) thus becomes

$$-\frac{\partial \bar{V}}{\partial t} - \frac{\partial a}{\partial t} - \left\langle \frac{\partial V_x}{\partial t}, \delta x \right\rangle - \tfrac{1}{2}\left\langle \delta x, \frac{\partial V_{xx}}{\partial t}\delta x \right\rangle = \min_{\delta u} [L(\bar{x}+\delta x, \bar{u}+\delta u; t)$$

$$+ \langle V_x + V_{xx}\delta x + \tfrac{1}{2} V_{xxx}\delta x\, \delta x, f(\bar{x}+\delta x, \bar{u}+\delta u; t)\rangle] \quad (1.4.13)$$

It should be noted that $\bar{V}+a$, V_x, and V_{xx} are functions of $\bar{x}$ and t; so

$$(d/dt)(\bar{V}+a) = (\partial/\partial t)(\bar{V}+a) + \langle V_x, f(\bar{x}, \bar{u}; t)\rangle \quad (1.4.14)$$

$$\dot{V}_x = \partial V_x/\partial t + V_{xx} f(\bar{x}, \bar{u}; t) \quad (1.4.15)$$

$$\dot{V}_{xx} = \partial V_{xx}/\partial t + \tfrac{1}{2} V_{xxx} f(\bar{x}, \bar{u}; t) + \tfrac{1}{2} f^T(\bar{x}, \bar{u}; t) V_{xxx} \quad (1.4.16)$$

Equations (1.4.10), (1.4.11) and (1.4.13) to (1.4.16) are used in Chapters 2 and 3 and in Appendix A to develop methods for determining the optimal control $u^o(t)$; $t \in [t_o, t_f]$ by successively improving the current nominal control $\bar{u}(t)$; $t \in [t_o, t_f]$.

1.4.2. Discrete-Time Systems

The principle of optimality yields the following difference equation for the optimal cost of a discrete-time system:

$$V_k^o(x_k) = \min_{u_k} [L_k(x_k, u_k) + V_{k+1}^o(x_{k+1})] \quad (1.4.17)$$

We assume that a nominal control sequence $\{\bar{u}_k: k = 1, ..., N-1\}$ and a nominal state sequence $\{\bar{x}_k: k = 1, ..., N\}$ are available. Writing x_k and u_k in terms of these nominal values yields

$$x_k = \bar{x}_k + \delta x_k \quad (1.4.18)$$

$$u_k = \bar{u}_k + \delta u_k \quad (1.4.19)$$

Substituting Equations (1.4.18) and (1.4.19) into Equation (1.4.17),

$$V_k^o(\bar{x}_k+\delta x_k) = \min_{\delta u_k} [L_k(\bar{x}_k+\delta x_k, \bar{u}_k+\delta u_k) + V_{k+1}^o(\bar{x}_{k+1}+\delta x_{k+1})] \quad (1.4.20)$$

We may expand $V_k^{\circ}(x_k)$ and $V_{k+1}^{\circ}(x_{k+1})$ about $\bar{x}_k$ and $\bar{x}_{k+1}$, respectively, (dropping the superscript for convenience):

$$V_k(\bar{x}_k+\delta x_k) = \bar{V}_k(\bar{x}_k) + a_k(\bar{x}_k) + \langle V_x{}^k, \delta x_k\rangle$$

$$+ \tfrac{1}{2}\langle \delta x_k, V_{xx}^k \delta x_k\rangle \qquad \text{plus higher-order terms} \qquad (1.4.21)$$

and

$$V_{k+1}(\bar{x}_{k+1}+\delta x_{k+1}) = \bar{V}_{k+1}(\bar{x}_{k+1}) + a_{k+1}(\bar{x}_{k+1}) + \langle V_x^{k+1}, \delta x_{k+1}\rangle$$

$$+ \tfrac{1}{2}\langle \delta x_{k+1}, V_{xx}^{k+1}\, \delta x_{k+1}\rangle \qquad \text{plus higher-order terms} \qquad (1.4.22)$$

Equation (1.1.7) then becomes

$$\bar{x}_{k+1}+\delta x_{k+1} = f_k(\bar{x}_k+\delta x_k, \bar{u}_k+\delta u_k); \qquad x_1 \quad \text{given} \qquad (1.4.23)$$

Equations (1.4.20) to (1.4.23) are used in Chapter 4 to develop algorithms for solving discrete-time deterministic control problems, and in Chapters 5 and 6 for discrete-time stochastic control problems.

1.5. SUMMARY

In this introductory chapter both continuous-time and discrete-time control problems were formulated and, in Section 1.2, some of the most important literature references pertaining to the theoretical study of these problems were indicated. Section 1.3 was devoted to brief descriptions of conventional numerical methods for solving optimal control problems. In Section 1.4 the notion of differential dynamic programming was introduced, from which we obtained equations that are obeyed by the local, quadratic expansion of the cost in the neighborhood of a nominal, nonoptimal trajectory. In Chapters 2, 3 and 4 and Appendix A, these equations will be used to devise new second-order and first-order algorithms for determining optimal control. Chapters 5 and 6 will describe the application of differential dynamic programming to stochastic optimal control problems.

---

† The quantities $\bar{V}_k$, $\bar{V}_{k+1}$, a_k, and a_{k+1} are defined in a similar way to $\bar{V}$ and a in Section 1.4.1.

References

1. G. A. Bliss, "Lectures on the Calculus of Variations," Chicago Univ. Press, Chicago, Illinois, 1959.
2. O. Bolza, "Lectures on the Calculus of Variations," Dover, New York, 1961.
3. F. A. Valentine, *in* "Contributions to the Calculus of Variations," p. 407. Chicago Univ. Press, Chicago, Illinois, 1937.
4. I. M. Gelfond and S. V. Fomin, "Calculus of Variations," Prentice-Hall, Englewood Cliffs, New Jersey, 1963.
5. S. F. Woods, "Advanced Calculus", p. 317. Ginn, Boston, Massachusetts, 1954.
6. C. Caratheodory, "Calculus of Variations and Partial Differential Equations of the First-Order," Holden-Day, San Francisco, California, 1965.
7. D. W. Bushaw, *in* "Contributions to the Theory of Non-Linear Oscillations," Vol. 2 Princeton Univ. Press, Princeton, New Jersey, 1958.
8. L. S. Pontryagin, V. G. Boltyanskii, R. V. Gamkrelidze, and E. F. Mischenko, "The Mathematical Theory of Optimal Processes", Wiley (Interscience), New York, 1962.
9. H. Halkin, *Arch. Rat. Mech. Anal.* **10**, 296 (1962).
10. H. Halkin, *J. Anal. Math.* **12**, 1 (1964).
11. H. Halkin, *in* "Non-Linear Differential Equations and Non-Linear Mechanics" (J. P. Lasalle, ed.), Academic Press, New York, 1963.
12. L. D. Berkovitz and S. E. Dreyfus, *J. Math. Anal. Appl.* **10**, 275 (1965).
13. S. E. Dreyfus, *J. Math. Anal. Appl.* **4**, 297 (1962).
14. S. E. Dreyfus, "Dynamic Programming and the Calculus of Variations," Academic Press, New York, 1965.
15. S. S. L. Chang, *Automatica* **1**, 55 (1963).
16. J. L. Speyer, Ph.D. Thesis, Div. Eng. and Appl. Phys., Harvard Univ., Cambridge, Massachusetts, 1968.
17. R. Bellman, "Dynamic Programming," Princeton Univ. Press, Princeton, New Jersey, 1957.
18. R. Bellman and S. E. Dreyfus, "Applied Dynamic Programming," Princeton Univ. Press, Princeton, New Jersey, 1962.
19. R. Bellman and R. Kalaba, *in* "Computing Methods in Optimization Problems" (A. V. Balakrishnan and L. W. Neustadt, eds.), p. 135. Academic Press, New York, 1964.
20. S. E. Dreyfus, *J. Math. Anal. Appl.* **1**, 228 (1960).
21. R. E. Kalman, *Bol. Soc. Mat. Mex.* **5**, 102 (1960).
22. R. E. Kalman, *in* "Mathematical Optimization Technique" (R. Bellman, ed.), California Univ. Press, Berkeley, California, 1963.
23. H. Halkin, *SIAM Rev.* **8**, 547 (October 1966). [Review of Dreyfus [14].]
24. H. Halkin, in "Advances in Control Systems" (C.T. Leondes ed.) p 173, Academic Press, New York, 1964.
25. H. J. Kelley, *in* "Optimization Techniques" (G. Leitman, ed.), Academic Press, New York, 1962.
26. A. E. Bryson and W. Denham, *J. Appl. Mech.* **29**, 247 (1962).
27. R. E. Larson, *IEEE Trans. Auto. Control* **AC 10**, 135 (1965).
28. S. Shapiro, Ph.D. Thesis, Univ. of London, England, 1965.
29. A. T. Fuller, *J. Electron. Control* **15**, 63 (1963).
30. A. T. Fuller, *J. Electron. Control* **17**, 283 (1964).
31. J. V. Breakwell and Yu-Chi Ho, *Intern. J. Eng. Sci.* **2**, 565 (1965).
32. M. R. Hestenes, *in* "Computing Methods in Optimization Problems" (A. V. Balakrishnan and L. W. Neustadt, eds.), p. 1. Academic Press, New York, 1964.

33. E. Lee, *J. SIAM Control* **1** 241, (1963).
34. L. Markus and E. Lee, *Trans. ASME: J. Basic Engineering* **84**, 13 (1962).
35. L. D. Berkovitz, *J. Math. Anal. Appl.* **3**, 145 (1961).
36. R. Bellman, I. Glicksberg, and O. Gross, *Quart. Appl. Math.* **14**, 11 (1956).
37. J. P. Lasalle, *in* "Contributions to the Theory of Non-Linear Oscillations," p. 1. Princeton Univ. Press, Princeton, New Jersey, 1960.
38. M. D. Levine, *Automatica* **3**, 203 (1966).
39. M. D. Levine, *Intern. J. Control* **3**, 39 (1966).
40. J. V. Breakwell, J. L. Speyer, and A. E. Bryson, *J. SIAM Control* **11**, 193 (1963).
41. A. H. Jazwinski, *Joint AIAA-IMS-SIAM-ONR Symp. on Control and Systems Optimization, U.S. Naval Postgraduate School, Monterey, California, 1964.*
42. L. W. Neustadt, *J. Math. Anal. Appl.* **1**, 484 (1960).
43. L. W. Neustadt, *Proc. 2nd Congr. Intern. Fed. Auto. Control, Basle, 1963*, p. 283.
44. B. Paiewonsky, *in* "International Symposium on Nonlinear Differential Equations and Nonlinear Mechanics" (J. P. LaSalle and S. Lefschetz, eds.), p.333 Academic Press, New York, 1963.
45. H. K. Knudsen, *IEEE Trans. Auto. Control* **AC-9**, 23 (1964).
46. R. E. Kalman, *Proc. IBM Sci. Comput. Symp. on Control Theory and Appl., Thomas. J. Watson Res. Center, Yorktown Heights, New York, 1964.*
47. C. H. Schley and I. Lee, *IEEE Trans. Auto. Control* **AC-12**, 139 (1967).
48. L. S. Lasdon, S. K. Mitter, and A. D. Waren, *IEEE Trans. Auto. Control*, **AC-12**, 132 (1967).
49. J. Allwright, Ph.D. Thesis, Univ. of London, England, 1969.
50. C. W. Merriam, III, "Optimization Theory and the Design of Feedback Controls, McGraw Hill, New York, 1964.
51. S. K. Mitter, *Automatica*, **3**, 135 (1966).
52. S. R. McReynolds and A. E. Bryson, *Proc. 6th Joint Auto. Control Conf., Troy, New York, 1965*, p. 551.
53. H. J. Kelley, *AIAA Astrodynamics Specialists Conf., Yale Univ. August 1963.*
54. T. E. Bullock and G. F. Franklin, *IEEE Trans. Auto. Control*, **AC-12**, 666 (1967).
55. H. Halkin, *in* "Computing Methods in Optimization Problems" (A. V. Balakrishnan and L. W. Neustadt, eds.) p. 211. Academic Press, New York, 1964.
56. D. Q. Mayne, *Intern. J. Control* **3**, 85 (1966).
57. B. Bernholtz and G. Graham, *AIEE Summer General Meeting, Atlanta, 1960, Part I; AIEE Summer General Meeting, Ithaca, 1961, Part II; AIEE Fall General Meeting, Chicago, 1964, Part III.*
58. D. H. Jacobson, Ph.D. Thesis, Univ. of London, England, 1967.
59. D. H. Jacobson, *Intern. J. Control* **7**, 175 (1968).
60. S. R. McReynolds, *J. Math. Anal. Appl.* **19**, 565 (1967).
61. D. H. Jacobson, *J. Opt. Theory Appl.* **2**, 411 (1968).
62. D. H. Jacobson, *IEEE Trans. Auto. Control*, **AC-13**, 661 (1968).
63. D. H. Jacobson and D. Q. Mayne, *Proc. 4th Congr. of Intern. Fed. Auto. Control, Warsaw, 1969.*
64. P. Dyer and S. R. McReynolds, *J. Math. Anal. Appl.* **23**, 585 (1968).
65. S. B. Gershwin and D. H. Jacobson, Harvard Univ. Tech. Rep. **TR 566** (August 1968).
66. D. Q. Mayne, Ph.D. Thesis, Univ. of London, England, 1967.
67. D. Q. Mayne, *Proc. 2nd IFAC Symp. on Theory of Self-Adaptive Control Systems, 1965.*
68. D. Q. Mayne, *J. Inst. Math. Appl.* **3**, 46 (1967).
69. J. H. Westcott, D. Q. Mayne, G. F. Bryant, and S. K. Mitter, *Proc. 3rd IFAC Congr., London, 1966.*

Chapter 2

NEW ALGORITHMS FOR THE SOLUTION OF A CLASS OF CONTINUOUS-TIME CONTROL PROBLEMS

2.1. INTRODUCTION

In this chapter we shall be concerned with the derivation of new second-order and first-order algorithms for determining optimal control of a class of continuous-time systems. The class of systems under consideration is the one in which the optimal control, as a function of time, is continuous. This is the most important and most widely studied optimal control problem because of the many physical control systems of interest that exhibit this property and, because, the absence of discontinuities† in its solution makes it amenable to analysis.

2.2. A NEW SECOND-ORDER ALGORITHM FOR UNCONSTRAINED PROBLEMS‡

2.2.1. The Derivation

We shall assume that the problem is formulated as in Section 1.1.1, but that all the constraints (1.1.3) to (1.1.6) are absent; later sections of this chapter treat constrained problems.

Equation (1.4.13), which is satisfied to second-order by the cost function in the neighborhood of a nominal trajectory, is repeated here, for convenience:

$$-\frac{\partial \bar{V}}{\partial t} - \frac{\partial a}{\partial t} - \left\langle \frac{\partial V_x}{\partial t}, \delta x \right\rangle - \tfrac{1}{2}\left\langle \delta x, \frac{\partial V_{xx}}{\partial t}\,\delta x \right\rangle = \min_{\delta u}\,[L(\bar{x}+\delta x, \bar{u}+\delta u; t)$$

$$+ \langle V_x + V_{xx}\,\delta x + \tfrac{1}{2} V_{xxx}\,\delta x\,\delta x, f(\bar{x}+\delta x, \bar{u}+\delta u; t)\rangle] \tag{2.2.1}$$

† Bang-bang control problems, in which the control function is piecewise continuous, are considered in Chapter 3.

‡ See Jacobson [1].

We now wish to use Equation (2.2.1) to devise algorithms for successively improving the nominal control function $\bar{u}(t)$; $t \in [t_o, t_f]$.

At any time $t \in [t_o, t_f]$, Equation (2.2.1) is valid locally with respect to δx but globally with respect to δu†; so, at time t, we may make large changes in δu.‡

Define

$$H(x, u, V_x; t) = L(x, u; t) + \langle V_x, f(x, u; t)\rangle \tag{2.2.2}$$

Equation (2.2.1) becomes

$$-\frac{\partial \bar{V}}{\partial t} - \frac{\partial a}{\partial t} - \left\langle \frac{\partial V_x}{\partial t}, \delta x \right\rangle - \tfrac{1}{2}\left\langle \delta x, \frac{\partial V_{xx}}{\partial t}\delta x \right\rangle = \min_{\delta u}\,[H(\bar{x}+\delta x, \bar{u}+\delta u, V_x; t)$$

$$+ \langle V_{xx}\delta x + \tfrac{1}{2} V_{xxx}\delta x\,\delta x, f(\bar{x}+\delta x, \bar{u}+\delta u; t)\rangle] \tag{2.2.3}$$

Let us consider Equation (2.2.3) at time t and let us first consider conditions for state $x = \bar{x}$. Equation (2.2.3) becomes

$$-(\partial \bar{V}/\partial t) - (\partial a/\partial t) = \min_{\delta u} H(\bar{x}, \bar{u}+\delta u, V_x; t) \tag{2.2.4}$$

Instead of using a second-order prediction of the minimizing δu, as is done by Mitter [2], McReynolds and Bryson [3], and Mayne [4], let us completely minimize H in Equation (2.2.4); this minimization may be done either analytically or, if necessary, numerically.

Assume that the minimizing u is

$$u^* = \bar{u} + \delta u^* \tag{2.2.5}$$

Now let us consider state $x = \bar{x}+\delta x$ at time t (i.e., reintroduce the variation δx). The minimizing control for $x = \bar{x}+\delta x$ will be

$$u = u^* + \delta u \tag{2.2.6}$$

where δu is yet to be determined.

† Because the nonlinearity of L and f with respect to δu is still preserved.

‡ Changing δu at time t has no effect on δx at time t; only $(d/dt)(\bar{x}+\delta x)$ is affected.

The right-hand side of Equation (2.2.3) becomes

$$\min_{\delta u} [H(\bar{x}+\delta x, u^*+\delta u, V_x; t) + \langle V_{xx}\delta x + \tfrac{1}{2} V_{xxx}\delta x\,\delta x, f(\bar{x}+\delta x, u^*+\delta u; t)\rangle]^\dagger \quad (2.2.7)$$

Expanding Expression (2.2.7) about $\bar{x}$, u^*, we obtain

$$\min_{\delta u} [H + \langle H_u, \delta u\rangle + \langle H_x + V_{xx} f, \delta x\rangle + \langle \delta u, (H_{ux} + f_u^T V_{xx})\delta x\rangle + \tfrac{1}{2}\langle \delta u, H_{uu}\delta u\rangle + \tfrac{1}{2}\langle \delta x, (H_{xx} + f_x^T V_{xx} + V_{xx} f_x)\delta x\rangle + \tfrac{1}{2}\langle V_{xxx}\delta x\,\delta x, f\rangle] \quad \text{plus higher-order terms} \quad (2.2.8)$$

All quantities in Expression (2.2.8) are evaluated at $\bar{x}$, u^*.

Now because u^* minimizes $H(\bar{x}, u, V_x; t)$ we have the necessary condition

$$H_u(\bar{x}, u^*, V_x; t) = 0 \quad (2.2.9)$$

In view of Equation (2.2.9), the terms in Expression (2.2.8) involving δu are

$$\langle \delta u, (H_{ux} + f_u^T V_{xx})\delta x\rangle + \tfrac{1}{2}\langle \delta u, H_{uu}\delta u\rangle \quad \text{plus higher-order terms} \quad (2.2.10)$$

If δu is of the same order as δx, then these terms will be quadratic in δx plus higher-order in δx. There is, therefore, no point in finding a relationship between δu and δx that is of order higher than linear, because terms higher than second-order in δx are neglected.‡ A relationship of the following form is, therefore, required:

$$\delta u = \beta\,\delta x \quad (2.2.11)$$

where β is chosen to minimize the contents of the square brackets in Expression (2.2.8).

A necessary condition for minimality is obtained by differentiating Expression (2.2.8) with respect to δu, and equating to zero:

$$H_u + H_{uu}\delta u + (H_{ux} + f_u^T V_{xx})\delta x \quad \text{plus higher-order terms} = 0 \quad (2.2.12)$$

† Notice that now δu is measured with respect to u^*.

‡ Because the left-hand side of Equation (2.2.3) is expanded to second-order only.

From Equation (2.2.9), $H_u = 0$. Substituting Equation (2.2.11) into Equation (2.2.12), we obtain

$$H_{uu}\beta\,\delta x + (H_{ux} + f_u^T V_{xx})\,\delta x \qquad \text{plus higher-order terms} = 0 \quad (2.2.13)$$

Equation (2.2.13) is required to hold for all δx sufficiently small, so we may equate coefficients of δx to zero, to obtain:

$$\beta = -H_{uu}^{-1}(H_{ux} + f_u^T V_{xx}) \qquad (2.2.14)$$

Quantities in Equation (2.2.14) are evaluated at $\bar{x}$, u^*. This β is the optimal, local linear feedback controller that maintains the necessary condition of optimality,

$$H_u(\bar{x} + \delta x, u^* + \delta u, V_x + V_{xx}\,\delta x; t) = 0 \qquad (2.2.15)$$

for δx sufficiently small.

Substituting Equation (2.2.11) into Equation (2.2.8) and neglecting terms higher than second-order in δx, we obtain

$$H + \langle H_x + V_{xx} f + \beta^T H_u, \delta x\rangle$$
$$+ \tfrac{1}{2}\langle \delta x, (H_{xx} + f_x^T V_{xx} + V_{xx} f_x - \beta^T H_{uu}\beta)\,\delta x\rangle + \tfrac{1}{2}\langle V_{xxx}\,\delta x\,\delta x, f\rangle \qquad (2.2.16)$$

Expression (2.2.16) equals the left-hand side of Equation (2.2.3). Since equality holds for all δx sufficiently small, the coefficients of like powers of δx may be equated to obtain

$$-(\partial \bar{V}/\partial t) - (\partial a/\partial t) = H$$

$$-(\partial V_x/\partial t) = H_x + \beta^T H_u + V_{xx} f \qquad (2.2.17)$$

$$-(\partial V_{xx}/\partial t) = H_{xx} + f_x^T V_{xx} + V_{xx} f_x - (H_{ux} + f_u^T V_{xx})^T H_{uu}^{-1} (H_{ux} + f_u^T V_{xx})$$
$$+ \tfrac{1}{2} V_{xxx} f + \tfrac{1}{2} f^T V_{xxx}$$

Using Equations (1.4.14) to (1.4.16), and noting that $-\dot{\bar{V}} = L(\bar{x}, \bar{u}; t)$, we obtain, from Equations (2.2.17),

$$-\dot{a} = H - H(\bar{x}, \bar{u}, V_x; t) \qquad (2.2.18)$$

$$-\dot{V}_x = H_x + \beta^T H_u^{=0} + V_{xx}(f - f(\bar{x}, \bar{u}; t)) \qquad (2.2.19)^\dagger$$

† The symbolism $H_u^{=0}$ indicates that $H_u(\bar{x}, u^*, V_x; t) = 0$.

$$-\dot{V}_{xx} = H_{xx} + f_x{}^T V_{xx} + V_{xx} f_x - (H_{ux} + f_u{}^T V_{xx})^T H_{uu}^{-1}(H_{ux} + f_u{}^T V_{xx})$$

$$+\tfrac{1}{2} V_{xxx}(f - f(\bar{x}, \bar{u}; t)) + \tfrac{1}{2}(f - f(\bar{x}, \bar{u}; t))^T V_{xxx} \tag{2.2.20}$$

Unless otherwise stated, all quantities are evaluated at $\bar{x}$, u^*.

$$\text{At} \quad t = t_f; \qquad V(\bar{x}; t_f) = F(\bar{x}(t_f); t_f) \tag{2.2.21}$$

whence

$$a(t_f) = 0 \tag{2.2.22}$$

$$V_x(t_f) = F_x(\bar{x}(t_f); t_f) \tag{2.2.23}$$

$$V_{xx}(t_f) = F_{xx}(\bar{x}(t_f); t_f) \tag{2.2.24}$$

Equations (2.2.22) to (2.2.24) are boundary conditions for the differential Equations (2.2.18) to (2.2.20). These equations are similar to those obtained by Mitter [2], McReynolds and Bryson [3], Mayne [4], McReynolds [5], and Bullock and Franklin [6]. Important differences are that the equations are evaluated at $\bar{x}$, u^* and *not* at $\bar{x}$, $\bar{u}$, and that V_{xxx} terms are present in (2.2.20).

The new control to be applied to the system is

$$u = \bar{u} + \delta u^* + \beta\,\delta x = u^* + \beta\,\delta x \tag{2.2.25}$$

This theory assumes that the δx, generated by Equation (1.4.4) on application of Equation (2.2.25), will be small enough to justify the second-order expansions used for V, L, and f. If δx becomes too large, one cannot simply place a scale factor ε $(0 < \varepsilon \leqslant 1)$ in front of δu^* as is done in [2], [3], and [4] and Appendix A, because H is often nonconvex in u which precludes this linear type of interpolation between $\bar{u}$ and $\bar{u} + \delta u^*$. (In the LQP problem there is no difficulty because immediate application of Equation (2.2.25) yields the optimal solution.)

Note that the differential equations (2.2.18) to (2.2.20) cannot be integrated owing to the presence of $V_{xxx}\Delta f$ and $\Delta f^T V_{xxx}$.† Let us neglect these terms and solve Equations (2.2.18) to (2.2.20); this yields $u^*(\tau)$ and $\beta(\tau)$. If the control policy

$$u(\tau) = u^*(\tau) + \beta(\tau)\,\delta x(\tau); \qquad \tau \in [t, t_f] \tag{2.2.26}$$

† $\Delta f = f(\bar{x}, u^*, t) - f(\bar{x}, \bar{u}, t)$. On an optimal trajectory $\Delta f = 0$.

is applied to the dynamic system, then it is easy to show that the equation

$$-\frac{\partial \bar{V}}{\partial t} - \frac{\partial \hat{a}}{\partial t} - \left\langle \frac{\partial \hat{V}_x}{\partial t}, \delta x \right\rangle - \tfrac{1}{2}\left\langle \delta x, \frac{\partial \hat{V}_{xx}}{\partial t} \delta x \right\rangle = H(\bar{x}, u^*, \hat{V}_x, t)$$

$$+ \langle H_u, \beta\,\delta x\rangle + \langle H_x + \hat{V}_{xx} f, \delta x\rangle + \langle \beta\,\delta x, (H_{ux} + f_u{}^T \hat{V}_{xx})\,\delta x\rangle$$

$$+ \langle \beta\,\delta x, H_{uu}\,\beta\,\delta x\rangle$$

$$+ \tfrac{1}{2}\langle \delta x, (H_{xx} + f_x{}^T \hat{V}_{xx} + \hat{V}_{xx} f_x)\,\delta x\rangle + \tfrac{1}{2}\langle \hat{V}_{xxx}\,\delta x\,\delta x, f\rangle \tag{2.2.27}$$

holds for δx sufficiently small. Here, the caret signifies that the control policy (2.2.26) is being used. All quantities in Equation (2.2.27) are evaluated at $\bar{x}$, u^*, $\hat{V}_x$. From Equation (2.2.27) and (1.4.14) to (1.4.16), and noting that $-\dot{\bar{V}} = L(\bar{x}, \bar{u}, t)$, we obtain

$$-\dot{\hat{a}} = H - H(\bar{x}, \bar{u}, \hat{V}_x, t)$$

$$-\dot{\hat{V}}_x = H_x + \beta^T H_u + \hat{V}_{xx}(f - f(\bar{x}, \bar{u}, t)) \tag{2.2.28}$$

$$-\dot{\hat{V}}_{xx} = H_{xx} + f_x{}^T \hat{V}_{xx} + \hat{V}_{xx} f_x + \beta^T (H_{ux} + f_u{}^T \hat{V}_{xx}) + (H_{ux} + f_u{}^T \hat{V}_{xx})^T \beta$$

$$+ \beta^T H_{uu} \beta + \tfrac{1}{2}\hat{V}_{xxx}(f - f(\bar{x}, \bar{u}, t)) + \tfrac{1}{2}(f - f(\bar{x}, \bar{u}, t))^T \hat{V}_{xxx}$$

The quantity $\hat{a}(t)$ is the predicted change in cost given that the control policy

$$u(\tau) = u^*(\tau) + \beta(\tau)\,\delta x(\tau); \qquad \tau \in [t, t_f] \tag{2.2.29}$$

is applied. Note that if the $\hat{V}_{xxx}$ terms are neglected in Equations (2.2.28), then Equations (2.2.28) are equivalent to Equations (2.2.18) to (2.2.20) with V_{xxx} terms neglected. Let us refer to the solutions of Equations (2.2.18) to (2.2.20), with V_{xxx} terms neglected, as $\tilde{a}$, $\tilde{V}_x$ and $\tilde{V}_{xx}$.† Since Equations (2.2.28) are linear, it is clear that neglect of the $\hat{V}_{xxx}$ terms introduces an error $\Delta\hat{V}_{xx}(\tau)$ in $\hat{V}_{xx}(\tau)$ of order‡

$$\int_{t_f}^{\tau} |u^*(t_1) - \bar{u}(t_1)|\, dt_1 \tag{2.2.30}$$

† β is thus equal to $-H_{uu}^{-1}(\bar{x}, u^*, \tilde{V}_x, t)\,[H_{ux}(\bar{x}, u^*, \tilde{V}_x, t) + f_u{}^T(\bar{x}, u^*, t)\,\tilde{V}_{xx}]$.

‡ Assuming that f and L satisfy a uniform Lipschitz condition with respect to changes in control u and that $\hat{V}_{xxx}$ is bounded. We define $|u| \triangleq \sum_{i=1}^{m} |u_i|$.

The error $\Delta \hat{V}_x(\tau)$ in $\hat{V}_x(\tau)$ is of order

$$\int_{t_f}^{\tau} \int_{t_f}^{t_2} |u^*(t_1) - \bar{u}(t_1)|\, dt_1\, |u^*(t_2) - \bar{u}(t_2)|\, dt_2 \tag{2.2.31}$$

and finally the error† $\Delta\hat{a}(t)$ in $\hat{a}(t)$ is of order

$$\int_{t_f}^{t} \int_{t_f}^{t_3} \int_{t_f}^{t_2} |u^*(t_1) - \bar{u}(t_1)|\, dt_1\, |u^*(t_2) - \bar{u}(t_2)|\, dt_2\, |u^*(t_3) - \bar{u}(t_3)|\, dt_3 \tag{2.2.32}$$

From Equation (2.2.28) it is clear that $\hat{a}(t)$ is of order

$$\int_{t_f}^{t} |u^*(t_4) - \bar{u}(t_4)|\, dt_4 \tag{2.2.33}$$

For t sufficiently close to t_f, i.e., $(t_f - t)$ of order ε, the error $\Delta\hat{a}(t)$ is of order ε^3, while $\hat{a}(t)$ is of order ε. If $|u^* - \bar{u}|$ is of order ε, then $\Delta\hat{a}(t)$ is of order ε^3 and $\hat{a}(t)$ is of order ε regardless of the value of t. In either case the result is that $\tilde{a}(t)$ and $\hat{a}(t)$ differ by a quantity of order ε^3; so an estimate of $\hat{a}(t)$, correct to second-order, is provided by $\tilde{a}(t)$. As stated prior to Equation (2.2.30), $\tilde{a}(t)$ is the $a(t)$ obtained from Equations (2.2.18) to (2.2.20) with V_{xxx} terms neglected. The upshot of the argument is that $\tilde{a}(t)$ is an estimate, correct to second-order, of the predicted change in cost produced by the control policy (2.2.26).

In the next section we describe a method of implementing the new control, Equation (2.2.26), so that the δx produced is small enough, and a reduction in cost is achieved at each iteration of the algorithm. We shall, for simplicity, write a, V_x, and V_{xx} without the tilde, but it must be remembered that these quantities have been obtained by neglecting V_{xxx} terms in Equation (2.2.20). The quantity $a(t)$ will be referred to as the predicted change in cost, though it is really an estimate.

(Note that in [1] the $V_{xxx}\delta x \delta x$ term in Equation (1.4.11) is neglected, *a priori*, without considering the above error analysis. The justification is that, for δx sufficiently small, Equations (1.4.10) and (1.4.11)—with $V_{xxx}\delta x \delta x$ neglected—form a model of the cost surface, in the neighborhood of $\bar{x}$, that is fully accurate to second-order in δx. The *a priori* neglect of the terms results in the same valid algorithms, but it takes an analysis of this type to demonstrate clearly that the error in the predicted change in cost, incurred by neglecting V_{xxx} terms, is third-order.)

† That is, $\hat{a}(t)$ and $\tilde{a}(t)$ differ by $\Delta\hat{a}(t)$.

2.2.2. A New "Step Size Adjustment Method"

Substituting Equation (2.2.25) into Equation (1.4.4), we obtain

$$(d/dt)(\bar{x}+\delta x) = f(\bar{x}+\delta x, u^* + \beta\, \delta x, t); \qquad \bar{x}(t_o) + \delta x(t_o) = x_o \tag{2.2.34}$$

Because $\delta x(t_o) = 0$, the δx produced by Equation (2.2.34) is owing to the driving action of $\delta u^* = u^* - \bar{u}$. The size of δx can be restrained and a reduction in cost can be obtained by altering the time interval over which Equation (2.2.34) is integrated.

Consider the time interval $[t_1, t_f]$ where $t_o \leqslant t_1 < t_f$. Assume that we run along the nominal trajectory $\bar{x}(t)$ from t_o to t_1. At time $t = t_1$, $\bar{x}(t_1) + \delta x(t_1) = \bar{x}(t_1)$, because we have followed along the nominal trajectory from t_o to t_1. Now consider integrating Equation (2.2.34) over the time interval $[t_1, t_f]$. If $t_1 < t_f$ and $[t_1, t_f]$ is small, then δx, produced by Equation (2.2.34), in this interval will be small, even for large δu^*, since there is very little time over which to integrate the differential equation:

$$(d/dt)(\bar{x}+\delta x) = f(\bar{x}+\delta x, u^* + \beta\, \delta x; t); \qquad \bar{x}(t_1) + \delta x(t_1) = \bar{x}(t_1) \tag{2.2.35}$$

By choosing t_1 near t_f, we can force δx to be as small as we please. Also, $\Delta a(t_1)$ is made negligible in comparison to $a(t_1)$.

The above description is summarized in the following statement: There exists a t_1, sufficiently close to t_f, in the range $t_o \leqslant t_1 < t_f$, such that if the nominal trajectory is followed from t_o to t_1 and then Equation (2.2.35) is integrated from t_1 to t_f,† the δx produced by Equation (2.2.35) in the interval $[t_1, t_f]$ will be small enough for the second-order expansions of V, L, and f to be valid, and the true predicted reduction in cost will be well described by $a(t_1)$.

The question "How does one choose t_1?" must be answered: Recall that $|a(\bar{x}; t_1)| = |\int_{t_f}^{t_1} [H - H(\bar{x}, \bar{u}, V_x; t)]\, dt|$ is the predicted improvement in cost when starting at the point $\bar{x}(t_1)$; t_1 and using

$$u(\tau) = u^*(\tau) + \beta(\tau)\delta x(\tau); \qquad \tau \in [t_1, t_f].$$

Assume for the moment that t_1 is set equal to t_o (i.e., consider the whole time interval $[t_o, t_f]$). Integrate Equation (2.2.35) and calculate the cost $V(\bar{x}; t_o)$. The actual improvement in cost is

$$\Delta V = \bar{V}(\bar{x}; t_o) - V(\bar{x}; t_o) \tag{2.2.36}$$

† That is, we effectively concatenate the old control $\bar{u}(t)$; $t \in [t_o\ t_1)$, with the new control $u^*(t) + \beta(t)\delta x(t)$; $t \in [t_1, t_f]$.

If this actual improvement in cost is "near" the predicted value $|a(\bar{x}; t_1)|$, then the new trajectory is considered to be satisfactory.

It is convenient to define "near" in the following way: If the following inequality is satisfied, ΔV is considered to be near $|a(\bar{x}; t_1)|$:

$$\Delta V/|a(\bar{x}; t_1)| > C; \qquad C \geqslant 0 \tag{2.2.37}$$

In practice C is set, say, as 0.5. There are no hard and fast rules for setting C; certainly it should be greater than or equal to zero, because a negative ΔV is inadmissible. C should not be greater than unity because one should not expect improvements in cost greater than predicted. Moreover, C should be somewhat less than unity so that any decisions based on Inequality (2.2.37) are not influenced by round-off errors in the computations.

If Inequality (2.2.37) is satisfied with $t_1 = t_o$, all is well, and the next iteration of the main algorithm may be begun with the knowledge that a reasonable reduction in cost of ΔV has been made. If Inequality (2.2.37) is not satisfied, then set

$$t_1 = (t_f - t_o)/2 + t_o = t_{o1} \tag{2.2.38}$$

The above procedure is repeated with this t_1, and Inequality (2.2.37) is checked again (with the new ΔV and t_1). If it is satisfied, then the next iteration is begun; if not, then set

$$t_1 = (t_f - t_{o1})/2 + t_{o1} = t_{o2} \tag{2.2.39}$$

and repeat again.

Subdividing $[t_o, t_f]$ in this way, there will come a time t_1 when Inequality (2.2.37) is satisfied. In general,

$$t_1 = (t_f - t_{or})/2 + t_{or} = t_{or+1} \tag{2.2.40}$$

where $r = 0, 1, \ldots$, and $t_{oo} = 2t_o - t_f$.

Notice that the new nominal trajectory will sometimes have a corner at t_1 because $\bar{u}(t_1)$ may be different from $u^*(t_1)$. This introduces no difficulty provided the integration routine being used is capable of handling differential equations with discontinuous right-hand sides.†

It may happen that the nominal trajectory $\bar{x}(t)$ is optimal on an interval $[t_2, t_f]$; $t_2 \in [t_o, t_f]$ but is nonoptimal on the interval $[t_o, t_f]$. If t_1 is being chosen in the manner outlined above, then a trial t_1 may fall in the interval

† For example, fourth-order Runge-Kutta method.

$[t_2, t_f]$. The δx generated in the interval $[t_1, t_f]$ would then be zero, since $u^*(t) = \bar{u}(t) = u^o(t)$; $t \in [t_2, t_f]$, and no reduction in cost would occur. One must ensure, therefore, that t_1 will never fall in $[t_2, t_f]$. This condition is ensured easily in the following way: At $t = t_f$, $a(\bar{x}; t) = 0$. When integrating the backward equations, monitor $|a(\bar{x}; t)|$. Record the time t_{eff} when $|a(\bar{x}; t)|$ becomes different from zero.[†] The trajectory between t_{eff} and t_f satisfies a necessary condition of optimality, viz,

$$a(\bar{x}; t) = 0; \qquad t \in [t_{eff}, t_f] \tag{2.2.41}$$

If, on the forward run, a time $t_1 \neq t_o$ needs to be found, then the interval $[t_o, t_{eff}]$ is subdivided as described earlier, and *not* $[t_o, t_f]$. As the over-all trajectory becomes more and more optimal, from iteration to iteration, so $t_{eff} \to t_o$. Finally, on an optimal trajectory $|a(\bar{x}; t)| < \eta$; $t \in [t_o, t_f]$ and $t_{eff} = t_o$, and the computation is halted.

When programming algorithms on a digital computer, it is generally necessary to use a numerical integration routine to integrate the differential equations. This means that the interval $[t_o, t_f]$ is divided into $N-1$ time steps (i.e., t from 1 to N). The subdivision of $[t_o, t_{eff}]$ used for determining t_1 must now be done with respect to this discretized time scale; i.e., a time N_1 must be sought, $N_1 \in [1, N_{eff}]$, where N_1 is given by

$$N_1 = (N_{eff} - N_{or})/2 + N_{or} = N_{or+1} \tag{2.2.42}$$

and where $N_{oo} = 2 - N_{eff}$ and $r = 0, 1, \ldots$. Integer division is used in Equation (2.2.42) (r may be increased until $N_1 = N_{eff} - 1$; if $N_{eff} = 1$, then only $r = 0$ is used).

It should be appreciated that since there are a finite number, $N-1$, of discrete-time steps, this subdivision can only be done a finite number of times. The smallest possible time interval is $(t_f - t_o)/(N-1)$. It is clear that N must be large enough such that the δx produced during this basic time interval is "small enough." This restriction is a practical one brought about by the discrete-time nature of the digital computation.

When ΔV and $|a(\bar{x}; t_1)|$ are small, but greater than η, the criterion (2.2.37) may be too severe with $C = 0.5$, owing to round-off errors or N being too small; i.e., there may come a stage where Inequality (2.2.37) remains unsatisfied even when $N_1 = N_{eff} - 1$. If this happens set $C = 0.0$ and repeat the procedure for determining N_1. $C = 0.0$ is a much less stringent test because it asks only that $\Delta V > 0$. If once again Inequality (2.2.37) is unsatisfied, even when $N_1 = N_{eff} - 1$, then stop the computation, since no further reduction in cost is possible. This implies that N is not large enough

[†] Or, in practice, when it becomes greater than a small positive quantity η.

and hence $(t_f - t_o)/(N-1)$ is too large a basic time interval (of course if $N_{eff} = 1$, then the trajectory is optimal).

Usually the N needed for accurate integration of the differential equations is large enough; the contrary has been encountered only in some problems that are near singular, and hence extremely sensitive to changes in u. In these cases it may prove desirable to use a very simple integration routine (Euler) and a large number of steps, N.

SUMMARY OF THE STEP SIZE ADJUSTEMENT METHOD:

1. When integrating the backward equations (2.2.18) to (2.2.20),† Section 2.2.1, note N_{eff} the time at which $|a(\bar{x};t)|$ becomes greater than η (η is chosen from numerical stability considerations).

2. On the forward run, try applying $u = u^* + \beta\delta x$ on the whole time interval $[t_0, t_f]$. Check whether Inequality (2.2.37) is satisfied with $C = 0.5$. If it is satisfied then proceed to the next iteration of the main algorithm; if not go to 3).

3. Use Equation (2.2.42) and Inequality (2.2.37) to subdivide the interval $[1, N_{eff}]$ in order to find a time N_1 such that Inequality (2.2.37) is satisfied. If no such N_1 can be found with $C = 0.5$, then set $C = 0.0$ and try again. If still no N_1 can be found, halt the computation.

Owing to the fact that the "step-size adjustment method" produces a corner into the new nominal trajectory, we are in fact introducing the notion of strong variations.

We define a weak variation in the following way: if we have a nominal trajectory $x(t) = \bar{x}(t)$; $t \in [t_1, t_f]$ then $\delta x(t)$; $t \in [t_1, t_f]$ is a weak variation of $x(t)$ if

$$|\delta x(t)| \leqslant \varepsilon \tag{2.2.43}$$

$$|(d/dt)[\delta x(t)]| \leqslant \varepsilon \tag{2.2.44}$$

$\varepsilon > 0$; $t \in [t_1, t_f]$. A strong variation occurs when Condition (2.2.43) holds but Condition (2.2.44) is *not* satisfied.

Strong variations have been used in theoretical studies of optimal control problems. In fact, Pontryagin proved his minimum principle using such variations. To our knowledge this is the first time that the notion of strong variations has been introduced into second-order successive approximation procedures.‡

---

† With V_{xxx} terms neglected.

‡ Halkin allows strong variations in his method of convex ascent [7].

It will be seen that this new algorithm solves a much larger class of problems than the algorithms of Mitter [2], McReynolds and Bryson [3], Mayne [4], and McReynolds [5].

2.2.3. The Computational Procedure

1. Using a nominal control $\bar{u}(t)$; $t \in [t_o, t_f]$, run a nominal $\bar{x}(t)$ trajectory; calculate the nominal cost $\bar{V}(x_o; t_o)$.

2. Using boundary conditions, Equations (2.2.22) to (2.2.24), integrate† Equations (2.2.18) to (2.2.20) backward in time from t_f to t_o, all the while minimizing H with respect to u to obtain $u^*(t)$ and $\beta(t)$. Note the time N_{eff} when $|a(\bar{x}; t)|$ becomes greater than η (η a small positive quantity).

3. Apply $u(t) = u^*(t) + \beta(t)\delta x(t)$ over the whole time interval $[t_o, t_f]$ and compute the resulting, new x trajectory and corresponding cost $V(x_o; t_o)$. If Inequality (2.2.37) is satisfied, repeat Step 2 after replacing the old nominal trajectory by the new, improved one. If the criterion is not satisfied, apply the "step size adjustment method" of Section 2.2.2; when a satisfactory trajectory is obtained, proceed to Step 2.

4. At the end of Step 2, check the size of $|a(x_o; t_o)|$; stop if it is less than η.

One might think that minimizing $H(\bar{x}, u, V_x; t)$ with respect to u at each instant of time going backward from t_f to t_o is "too much work." In practice, however, it is quite straightforward. There are two cases to consider: (1) the minimization of H can be done analytically. In this case there are no difficulties. (2) the minimization of H must be done using a numerical method.

During the past 10 years, some very powerful hill-climbing techniques (minimizing methods) have emerged [8, 9]. In our opinion, the method described by Fletcher and Powell [9] is a considerable advance over older methods (Newton-Raphson, steepest descent). The function to be minimized and its first derivatives are calculated, and the inverse matrix of second derivatives is estimated from this data as the procedure progresses toward the minimum of the function. Thus, when the minimum is reached, an estimate of the inverse of the matrix of second derivatives evaluated at the minimum is available. This estimate is constructed in such a way that it is positive-definite.

Consider using the method described by Fletcher and Powell [9], abbreviated F and P, for minimizing H with respect to u. At $t = t_f$, we use F and P to minimize H. The procedure supplies us with an estimate of $H_{uu}^{-1}(\bar{x}, u^*, V_x; t_f)$ that we can use as our best estimate of $H_{uu}^{-1}(t_f - \Delta t)$ (we have dropped the other arguments for convenience). Note that if $u^*(t)$ and $H(t)$ are continuous, smooth functions of time, then H_{uu}^{-1} changes little in an interval Δt. This means that $H_{uu}^{-1}(t_f)$ is a good estimate of $H_{uu}^{-1}(t_f - \Delta t)$

† V_{xxx} terms are neglected.

and $u^*(t_f)$ is a good estimate of $u^*(t_f - \Delta t)$ so that, at $t_f - \Delta t$, the F and P routine has little difficulty in quickly attaining the minimum of $H(t_f - \Delta t)$. This argument holds also for any $t \in [t_o, t_f]$. We note that $\hat{H}_{uu}^{-1}(t)$, generated by F and P, is an estimate of H_{uu}^{-1} at u^*, t so that when we have finished minimizing $H(t)$, we may use $\hat{H}_{uu}^{-1}(t)$ to calculate $\beta(t)$. However, it is possible that the minimum of $H(t)$ is reached and the resulting $\hat{H}_{uu}^{-1}(t)$ is not a very good estimate of the true $H_{uu}^{-1}(t)$ evaluated at the minimum $u^*(t)$. The goodness of this estimate can be checked and updated readily in the following way: we have

$$\hat{H}_{uu}^{-1} H_{uu} = I + \varepsilon \tag{2.2.45}$$

where ε is a matrix that is a measure of the difference between $\hat{H}_{uu}^{-1}$ and H_{uu}^{-1}. Now if the elements of ε are large, then $\hat{H}_{uu}^{-1}$ is a poor estimate of H_{uu}^{-1}; thus the best we can do is to invert H_{uu} and use it for determining β. If the elements of ε are small, then from Equation (2.2.45),

$$\hat{H}_{uu}^{-1} = [I + \varepsilon] H_{uu}^{-1}$$

whence

$$H_{uu}^{-1} = [I - \varepsilon + \varepsilon^2/2!, \ldots] \hat{H}_{uu}^{-1}$$

If ε is quite small, then

$$H_{uu}^{-1} \doteqdot [I - \varepsilon] \hat{H}_{uu}^{-1} \tag{2.2.46}$$

Equations (2.2.45) and (2.2.45) may be used recursively to determine H_{uu}^{-1} from $\hat{H}_{uu}^{-1}$ and H_{uu}.

In many control problems, one finds that there are relatively few control variables, and so there is not much point in using Equations (2.2.45) and (2.2.46), since H_{uu} can be inverted easily. Equations (2.2.45) and (2.2.46) could be useful if there were more than, say, three controls.

2.2.4. Characteristics of the Algorithm

1. The procedure exhibits one step convergence on LQP problems.

2. In a neighborhood of the optimum, convergence is rapid for non-LQP problems because the second-order expansions represent the functions V, L, and f well, for small δx and δu.

3. The algorithm is very much more powerful than the existing methods [2-4] for the following reason: $H_{uu}^{-1}(\bar{x}, \bar{u}, V_x; t)$ is not required to be positive-definite along nonoptimal nominal trajectories. In this algorithm

$H(\bar{x}, u, V_x; t)$ is minimized with respect to u, and so it is required only that $H_{uu}^{-1}(\bar{x}, u, V_x; t)$ be positive-definite at the minimizing $u = u^*$; i.e., $H(\bar{x}, u, V_x; t)$ must be strictly convex only in the neighborhood of u^*. This is a much less restrictive requirement; thus the algorithm is capable of handling a larger class of non-LQP problems than the second-order or second-variation methods.

4. In some problems the solution of the Riccati equation becomes unbounded along some nominal trajectories, though along optimal trajectories, it always has a bounded solution. The new algorithm is able to compute optimal control for these problems, whereas the existing methods are not; this will be illustrated by Example 2.

5. If the δx produced by the new control is too large, as measured by the criterion (2.2.37), then the "step size adjustment routine" must be used. If the problem is very non-LQP, the routine will have to be used a number of times in order to determine t_1, which will be close to t_f. However, as $t_1 \to t_f$, Equation (2.2.35) is integrated over ever-decreasing time intervals $[t_1, t_f]$. This is in contrast with methods [2-4] where, in order to determine the scale factor ε, the $\dot{x}$ equation has to be integrated over the whole time interval $[t_0, t_f]$.† So, in non-LQP problems, it is likely that the new algorithm will use less computing time in determining t_1 than existing methods use in determining ε.

6. The algorithm requires the integration of n differential equations less than those required by Mitter [2] and McReynolds and Bryson [3], who integrate an n-vector differential equation $\dot{h}$, in addition to their equation for $\dot{\lambda}$.

2.2.5. A Computational Trick that Improves Convergence Rate

In the algorithm the new control is computed using

$$u(t) = u^*(t) + \beta(t)\,\delta x(t) \tag{2.2.47}$$

It can happen in non-LQP problems that $\beta(t)\,\delta x(t)$ becomes too large, and so invalidates the local expansions in δu. However, δx might still be small enough for

$$V_x(\bar{x}+\delta x; t) = V_x + V_{xx}\,\delta x \tag{2.2.48}$$

to be valid.

The following alternative can be used for computing $u(t)$: instead of storing $u^*(t)$ and $\beta(t)$, store $V_x(t)$ and $V_{xx}(t)$. Compute $u(t)$ directly by

† $u = \bar{u} + \varepsilon\,\delta u^* + \beta\,\delta x.$

minimizing $H(\bar{x}+\delta x, u, V_x + V_{xx}\delta x; t)$ with respect to u, either analytically or using F and P. In this way the radius of convergence of the algorithm may be increased.

2.2.6. Sufficient Conditions for a Reduction in Cost at Each Iteration

In order that the cost decrease at each iteration, for δx sufficiently small, $a(\bar{x}; t)$ must be less than zero. Sufficient conditions for $a(\bar{x}; t_1) < 0$ are that, for $t \in [t_o, t_f]$,

(1) $H(\bar{x}, u^*, V_x; t) < H(\bar{x}, \bar{u}, V_x; t), \qquad u^* \neq \bar{u}$

(2) $H_{uu}^{-1}(\bar{x}, u^*, V_x; t)$ is positive-definite

(3) The solutions of Equations (2.2.18) to (2.2.20) be bounded.

PROOF:

$$a(\bar{x}; t_1) = \int_{t_f}^{t_1} [H(\bar{x}, u^*, V_x; t) - H(\bar{x}, \bar{u}, V_x; t)]\, dt \tag{2.2.49}$$

For $a(\bar{x}; t_1) < 0$, it is clearly sufficient that

$$H(\bar{x}, u^*, V_x; t) < H(\bar{x}, \bar{u}, V_x; t), \quad u^* \neq \bar{u} \tag{2.2.50}$$

u^* is the control that minimizes H; so (2.2.50) is true for $u^* \neq \bar{u}$.

The quantities manipulated above must be bounded in magnitude for these conditions to be valid; so it is required that the solutions of Equations (2.2.18) to (2.2.20) be bounded (Jacobi or conjugate point condition, when $\bar{u} = u^o$).

2.2.7. A New First-Order Algorithm for Unconstrained Problems†

Assume that $V(\bar{x}+\delta x; t)$ is expanded to first-order only:

$$V(\bar{x}+\delta x; t) = \bar{V} + a + \langle V_x, \delta x \rangle \tag{2.2.51}$$

The following set of equations is easily obtained:‡

$$-\dot{a} = H - H(\bar{x}, \bar{u}, V_x; t); \qquad a(t_f) = 0 \tag{2.2.52}$$

$$-\dot{V}_x = H_x; \qquad V_x(t_f) = F_x(\bar{x}(t_f); t_f) \tag{2.2.53}$$

---

† See Jacobson [1].

‡ Where the term $V_{xx}\,\Delta f$ is neglected in Equation (2.2.53).

The quantities are evaluated at $\bar{x}, u^*; t$ unless otherwise stated; u^* is the control that minimizes $H(\bar{x}, u, V_x; t)$ with respect to u.

The new control that is applied is

$$u(t) = u^*(t), \qquad t \in [t_1, t_f] \tag{2.2.54}$$

and the "step size adjustment method" is used.

2.2.8. Characteristics of the First-Order Algorithm

1. The algorithm is fundamentally different from the gradient or first-variation method in that Equations (2.2.52) and (2.2.53) are integrated backward along $\bar{x}, u^*$, and the new control on the forward run is given by Equation (2.2.54).

2. The algorithm uses the step size adjustment routine described earlier. Since V is expanded to first-order, it might be necessary to repeatedly use the step size adjustment routine a number of times before an acceptable t_1 is found. However, the integration of the $\dot{x}$ equation is done over ever-decreasing time intervals $[t_1, t_f]$, and so it is likely that the method will be faster than the gradient method, where ε has to be chosen. Note also that $t_1 \in [t_o, t_f]$. In the gradient method ε is required to be greater than zero but no upper bound is available for it; this usually makes the choice of ε tricky.

3. Consider the problem:

$$\dot{x} = Ax + Bu \tag{2.2.55}$$

$$\min V = \int_{t_o}^{t_f} L(u; t)\, dt + \langle c, x(t_f)\rangle \tag{2.2.56}$$

If L is a strictly convex function of u, it can be shown that the first-order algorithm solves this problem in one step since V^0 is linear in x. This is not true of the gradient method where only a small change in control, $\delta u = -\varepsilon H_u$, is made at each iteration.

2.2.9. Computed Examples

The following computed examples serve to illustrate some of the advantages of the new second-order algorithm:

$$\dot{x} = -0.2x + 10 \tanh u; \qquad x(t_o) = 5.0 \tag{2.2.57}$$

Choose $u(t)$; $t \in [0, 0.5]$ to minimize

$$V(x_o; t_o) = \int_0^{0.5} (10x^2 + u^2)\,dt + 10x^2(t_f) \tag{2.2.58}$$

The problem, though simple, is a good illustrative one because along certain nonoptimal trajectories, $H_{uu}^{-1}(\bar{x}, \bar{u}, V_x; t)$ is not positive. For this problem,

$$H(x, u, V_x; t) = 10x^2 + u^2 + V_x(-0.2x + 10\tanh u) \tag{2.2.59}$$

where

$$H_u = 2u + 10V_x(1 - \tanh^2 u) \tag{2.2.60}$$

and

$$H_{uu} = 2 - 20V_x \tanh u(1 - \tanh^2 u) \tag{2.2.61}$$

The new second-order algorithm and those of Mitter [2], McReynolds and Bryson [3], and Mayne [4] were programmed. A fourth-order Runge-Kutta routine was used for the integration, and the interval [0, 0.5] was divided into 100 steps. From Equation (2.2.61), it is clear that there is no guarantee that H_{uu}^{-1} will be greater than zero for any nominal trajectory $\bar{x}, \bar{u}$; i.e., there is no guarantee the methods [2–5] will be successful.

The new algorithm requires only that $H_{uu}^{-1}(\bar{x}, u^*, V_x; t) > 0$. At $u = u^*$, one has, since $H_u(\bar{x}, u^*, V_x; t) = 0$, that

$$\begin{aligned} 1 - \tanh^2 u^* &= -2u^*/10V_x \qquad \text{if} \quad V_x \neq 0 \\ 1 - \tanh^2 u^* &= 1 \qquad\qquad\quad\;\; \text{if} \quad V_x = 0 \end{aligned} \tag{2.2.62}$$

Using Equation (2.2.62) in Equation (2.2.61),

$$H_{uu}(\bar{x}, u^*, V_x; t) = 2 + 4u^* \tanh u^* \tag{2.2.63}$$

$u^* \tanh u^* \geqslant 0$ for all u^*; so, from Equation (2.2.63), $H_{uu}(\bar{x}, u^*, V_x; t) > 0$ *regardless* of the nominal trajectory. The new algorithm, therefore, should not fail to solve this problem.

A nominal control $u(t) = +1$; $t \in [0, 0.5]$ was chosen and an attempt was made to use methods described by workers [2-5]. For this nominal control, $H_{uu}(\bar{x}, \bar{u}, V_x; t)$ turned out to be *negative* for $t \in [0, 0.5]$ and so the algorithms were unable to improve the trajectory.

Starting from the same nominal trajectory, the new algorithm was tried. By Equation (2.2.63), $H_{uu}(\bar{x}, u^*, V_x; t)$ remained positive on the interval [0, 0.5] (u^* was determined by quadratic prediction [10]). On the forward run, a reduction in cost was achieved and after two iterations, the optimal trajectory was reached. The cost was reduced from the nominal value of 886.0 to the optimal value of 41.6. The trajectory was considered optimal when $|a(x_0; t_0)|$, the predicted reduction in cost, was less than 0.1. Figure 2.1 shows the nominal and optimal control functions.

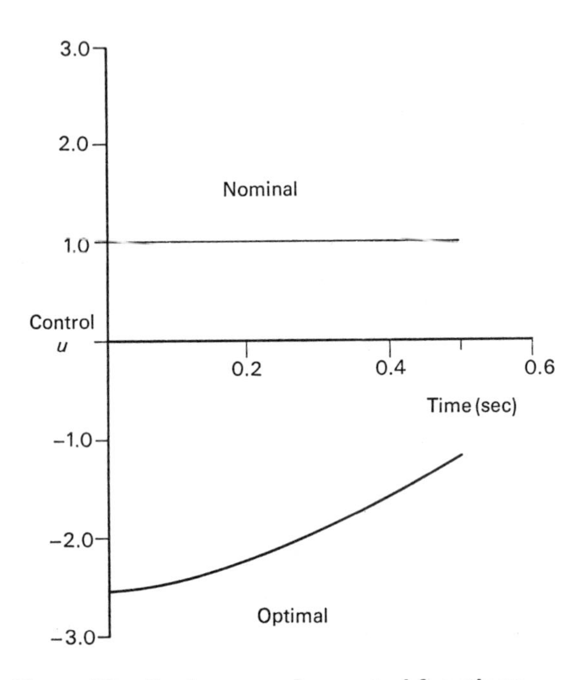

Figure 2.1. Scalar example: control functions.

This simple example illustrates the failure of the methods described [2-5] to find a solution to a control problem where the nominal trajectory is such that $H_{uu}^{-1}(\bar{x}, \bar{u}, V_x; t)$ is nonpositive-definite. The new algorithm, where $H(\bar{x}, u, V_x; t)$ is minimized with respect to u, easily finds the optimal trajectory.

It should be noted that when using the new algorithm, the "computational trick" of minimizing $H(\bar{x}+\delta x, u, V_x + V_{xx}\delta x; t)$ with respect to u was used on the forward run. The same problem was tried using Equation (2.2.47) to calculate the new control; four iterations were required to reach the optimum. This behavior is due to the fact that H is very nonlinear in u and so, $u = u^* + \beta \delta x$ is valid only for very small variations δu from u^*. An increased radius of convergence is thus obtained by choosing u by

$$\min_u H(\bar{x} + \delta x, u, V_x + V_{xx} \delta x; t).$$

2. *The Rayleigh equation.* In this example the solution of the Riccati equation becomes unbounded when integrating backward from t_f to t_o along some nominal trajectories. It is demonstrated that the new second-order algorithm is still able to achieve a reduction in cost at each iteration and moreover, reaches the optimal trajectory after 9 iterations. The methods described by Mitter [2], McReynolds and Bryson [3], Mayne [4], and McReynolds [5] fail to solve this problem.

Consider the following control problem:

$$\begin{aligned} \dot{x}_1 &= x_2 \quad ; & x_1(t_o) &= -5. \\ \dot{x}_2 &= -x_1 + 1.4 x_2 - 0.14 x_2^3 + 4u ; & x_2(t_o) &= -5. \end{aligned} \tag{2.2.64}$$

Find $u(t)$; $t \in [0, 2.5]$, to minimize

$$V = \int_0^{2.5} (x_1^2 + u^2)\, dt \tag{2.2.65}$$

$$H = x_1^2 + u^2 + V_{x_1} x_2 + V_{x_2}(-x_1 + 1.4 x_2 - 0.14 x_2^3 + 4u) \tag{2.2.66}$$

$$H_u = 2u + 4 V_{x_2}; \qquad \text{whence} \quad u^* = -2 V_{x_2} \tag{2.2.67}$$

$$H_{uu} = 2 > 0 \tag{2.2.68}$$

$$H_x = \begin{bmatrix} 2x_1 - V_{x_2} \\ V_{x_1} + 1.4 V_{x_2} - 0.42 x_2^2 V_{x_2} \end{bmatrix} \tag{2.2.69}$$

$$H_{ux} = 0 \tag{2.2.70}$$

$$H_{xx} = \begin{bmatrix} 2 & 0 \\ 0 & -0.84\, V_{x_2} x_2 \end{bmatrix} \tag{2.2.71}$$

An arbitrary nominal control of

$$\bar{u}(t) = -0.5; \qquad t \in [0,\ 2.5] \tag{2.2.72}$$

was chosen.

The fourth-order Runge-Kutta routine was used for the integrations and one-hundred integration steps were used. The methods previously described [2-5] were tried first. It was found that during the backward integration of the $\dot{V}_x$, $\dot{V}_{xx}$ equations, their solutions became unbounded. The integration step size was reduced by increasing the number of integration steps from 100 to 1000, but the same behavior persisted; this meant that the methods could not be used.

Since, for this problem, the $\dot{V}_x$ equation is linear in V_x, its solution can be unbounded only if V_{xx} becomes unbounded (i.e., if the solution of the Riccati equation becomes unbounded).

From Kalman [11], sufficient conditions for the boundedness of the Riccati solution are that

$$H_{xx} - H_{ux}^T H_{uu}^{-1} H_{ux} \quad \text{is positive-semidefinite}$$

$$H_{uu}^{-1} \quad \text{is positive-definite}$$

$$F_{xx}(\bar{x}(t_f); t_f) \quad \text{is positive-semidefinite}$$

The second and third conditions are satisfied since $H_{uu} = 2$, $F_{xx} = 0$; in addition, $H_{ux} = 0$.

It turns out that H_{xx}, given by Equation (2.2.70), is not positive-semi-definite all along the nominal trajectory (i.e., the above sufficiency conditions for the boundedness of the solution of the Riccati equation are not satisfied); this could account for the observed unboundedness of the V_{xx} solution.

The new algorithm was tried next; again the solutions of the backward equations became unbounded. However, using the step size adjustment method, a t_1 could be found; $t_b < t_1 < t_f$ (t_b being the time at which the solutions became unbounded) such that an improvement in cost resulted. As the method progressed from iteration to iteration, so t_b approached t_0. Finally, the optimal trajectory along which all the equations had bounded solutions was reached (i.e., the new second-order algorithm was entirely successful).

Figure 2.2 shows $V_{x_2}(t)$ for various iterations, illustrating how the time t_b moved backward to t_o as the optimum trajectory was approached. Figure 2.3 shows the cost V as a function of the iteration number.

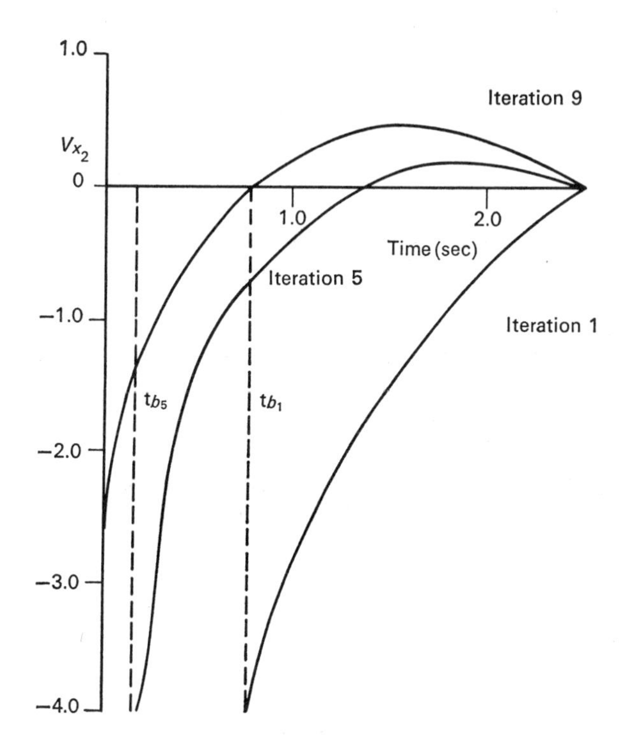

Figure 2.2. Rayleigh equation: V_{x_2} versus Time.

The above example shows that the new algorithm is more powerful than existing ones in the sense that the boundedness of the solutions of the backward equations is not required on the whole interval $[t_o, t_f]$, along non-optimal trajectories. It is required only that they have bounded solutions on the interval $[t_b, t_f]$; $t_b < t_f$ and that $x(t)$; $t \in [t_b, t_f]$ be nonoptimal unless $t_b < t_o$.

It should be clear from the above, that the conditions of Section 2.2.6 are "over sufficient."

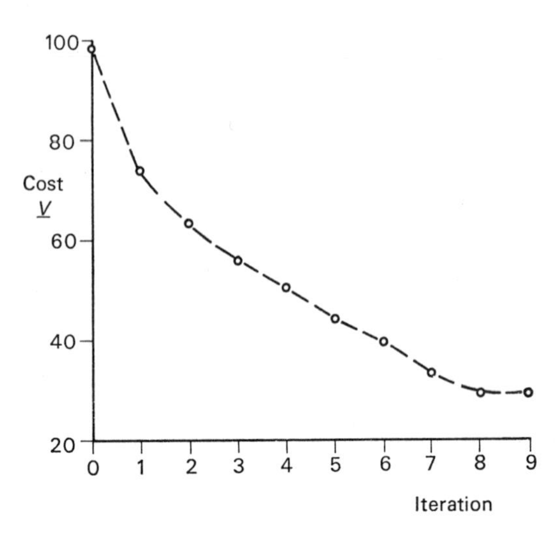

Figure 2.3. Rayleigh equation: cost versus iteration number.

2.3. A NEW SECOND-ORDER ALGORITHM FOR FIXED ENDPOINT PROBLEMS

2.3.1. The Derivation

In this section we shall consider the class of problems where the endpoint of the trajectory must obey the following equality:

$$\psi(x(t_f); t_f) = 0 \tag{2.3.1}$$

where ψ is an $s \leqslant n$-dimensional vector function. Here we shall assume that the final time t_f is given explicitly; the case where t_f is given implicitly is considered in Section 2.3.5.

The control problem is to choose $u(t)$; $t \in [t_o, t_f]$ in order to minimize the performance index, Equation (1.1.2), and to satisfy the constraints described by Equation (2.3.1).

We shall convert this constrained optimization problem into an unconstrained one by adjoining the constraints to the performance index using an s-dimensional, time invariant vector Lagrange multiplier b:

$$V(x_o, b; t_o) = \int^{t_f} L(x, u, t)\, dt + F(x(t_f); t_f) + \langle b, \psi(x(t_f); t_f)\rangle \tag{2.3.2}$$

The control problem is now to choose $u(t)$; $t \in [t_o, t_f]$ such that V, given by Equation (2.3.2), is minimized† and to choose b such that Equation (2.3.1) is satisfied.‡

Let us assume that we do not know the optimal control $u^o(t)$; $t \in [t_o, t_f]$ or the optimal multipliers b^o. We have only a nominal control $\bar{u}(t)$; $t \in [t_o, t_f]$ and a nominal set of multipliers $\bar{b}$. The nominal control produces a nominal trajectory via Equation (1.1.1), and Equation (2.3.2) yields a nominal cost $\bar{V}(x_o, \bar{b}; t_o)$.

In a similar fashion to that described in Section 1.4.1, let us expand $V(x, b; t)$ to second-order about $\bar{x}$, $\bar{b}$:

$$V(\bar{x}+\delta x, \bar{b}+\delta b; t) = \bar{V}(\bar{x}, \bar{b}; t) + a + \langle V_x, \delta x\rangle + \langle V_b, \delta b\rangle + \langle \delta b, V_{xb}^T \delta x\rangle + \tfrac{1}{2}\langle \delta x, V_{xx}\delta x\rangle + \tfrac{1}{2}\langle \delta b, V_{bb}\delta b\rangle \tag{2.3.3}$$

The Bellman equation in the neighborhood of the nominal trajectory becomes:§

$$-\frac{\partial \bar{V}}{\partial t} - \frac{\partial a}{\partial t} - \left\langle \frac{\partial V_x}{\partial t}, \delta x\right\rangle - \left\langle \frac{\partial V_b}{\partial t}, \delta b\right\rangle - \left\langle \delta b, \frac{\partial V_{xb}^T}{\partial t}\delta x\right\rangle - \tfrac{1}{2}\left\langle \delta x, \frac{\partial V_{xx}}{\partial t}\delta x\right\rangle - \tfrac{1}{2}\left\langle \delta b, \frac{\partial V_{bb}}{\partial t}\delta b\right\rangle$$

$$= \min_{\delta u}\left[L(\bar{x}+\delta x, \bar{u}+\delta u; t) + \langle V_x + V_{xx}\delta x + V_{xb}\delta b, f(\bar{x}+\delta x, \bar{u}+\delta u; t)\rangle\right] \tag{2.3.4}$$

† It is possible that this new cost functional (or performance index) does not possess a minimum but only a stationary point, with respect to the function u; we ignore this possibility.

‡ It turns out that V must be maximized with respect to b.

§ In this, and subsequent algorithms, terms $V_{xbb}\delta b\delta b$, $V_{xxb}\delta x\delta_b$, and $V_{xxx}\delta x\delta x$ are omitted *a priori*, for convenience, rather than carrying them throughout the derivation and then neglecting them at the end, as was done in Section 2.2.

Define

$$H(x, u, V_x; t) = L(x, u; t) + \langle V_x, f(x, u; t)\rangle \tag{2.3.5}$$

and consider the right-hand side of Equation (2.3.4), with δx and δb equal to zero:

$$\min_{\delta u} H(\bar{x}, \bar{u}+\delta u, V_x; t) \tag{2.3.6}$$

As in Section 2.2.1, we determine the δu^* that minimizes H, and so Expression (2.3.6) becomes

$$H(\bar{x}, u^*, V_x; t) \tag{2.3.7}$$

where

$$u^* = \bar{u} + \delta u^* \tag{2.3.8}$$

As in Section 2.2.1, we reintroduce the variations δx and δb, and also the δu required to maintain optimality:

$$\min_{\delta u} [H(\bar{x}+\delta x, u^*+\delta u, V_x; t) + \langle V_{xx}\delta x + V_{xb}\delta b, f(\bar{x}+\delta x, u^*+\delta u; t)\rangle] \tag{2.3.9}$$

Again following the approach used in Section 2.2.1, we obtain the linear relationship between δu and δx and δb:†

$$\delta u = \beta_1 \delta x + \beta_2 \delta b \tag{2.3.10}$$

where

$$\beta_1 = -H_{uu}^{-1}(H_{ux} + f_u{}^T V_{xx}) \tag{2.3.11}$$

$$\beta_2 = -H_{uu}^{-1} f_u{}^T V_{xb} \tag{2.3.12}$$

Substituting Equation (2.3.10) into Equation (2.3.9), neglecting terms of order higher than the second, and equating to the left-hand side of Equa-

† Details are not given here as the analysis is almost identical to that described in Section 2.2.1.

tion (2.3.4), we obtain:

$$
\begin{aligned}
-\dot{a} &= H - H(\bar{x}, \bar{u}, V_x; t) \\
-\dot{V}_x &= H_x + V_{xx}(f - f(\bar{x}, \bar{u}; t)) + \beta_1{}^T \overset{=0}{H_u} \\
-\dot{V}_b &= V_{xb}^T(f - f(\bar{x}, \bar{u}; t)) + \beta_2{}^T \overset{=0}{H_u} \\
-\dot{V}_{xb} &= (f_x{}^T + \beta_1{}^T f_u{}^T) V_{xb} \qquad\qquad (2.3.13) \\
-\dot{V}_{bb} &= -V_{xb}^T f_u H_{uu}^{-1} f_u{}^T V_{xb} \\
-\dot{V}_{xx} &= H_{xx} + f_x{}^T V_{xx} + V_{xx} f_x - (H_{ux} + f_u{}^T V_{xx})^T H_{uu}^{-1} (H_{ux} + f_u{}^T V_{xx})
\end{aligned}
$$

where all quantities are evaluated at $\bar{x}$, $\bar{b}$, and u^* unless otherwise stated.

The new control is

$$u(t) = u^*(t) + \beta_1(t)\delta x(t) + \beta_2(t)\delta b \tag{2.3.14}$$

At $t = t_f$, we have, from Equation (2.3.2), that

$$V(\bar{x}, \bar{b}; t_f) = F(\bar{x}(t_f); t_f) + \langle \bar{b}, \psi(\bar{x}(t_f); t_f) \rangle \tag{2.3.15}$$

whence

$$
\left.
\begin{aligned}
a(t_f) &= 0 \\
V_x(t_f) &= F_x(\bar{x}(t_f); t_f) + \psi_x{}^T(\bar{x}(t_f); t_f)\bar{b} \\
V_b(t_f) &= \psi(\bar{x}(t_f); t_f) \\
V_{xb}(t_f) &= \psi_x{}^T(\bar{x}(t_f); t_f) \\
V_{bb}(t_f) &= 0 \\
V_{xx}(t_f) &= F_{xx}(\bar{x}(t_f); t_f) + \bar{b}\psi_{xx}(\bar{x}(t_f); t_f)
\end{aligned}
\right\} \qquad (2.3.16)
$$

These are boundary conditions, for Equations (2.3.13) at $t = t_f$.

Let us consider now a computational procedure for improving the nominal control $\bar{u}(t)$; $t \in [t_o, t_f]$ and the nominal multipliers $\bar{b}$. First, let us set $b = \bar{b}$ and, using the procedure of Section 2.2.3, reduce $|a(x_o, \bar{b}; t_o)|$ to zero; i.e., let us solve the free endpoint problem with $b = \bar{b}$. Having reduced

$|a(x_o, \bar{b}; t_o)|$ to zero, we note from Equations (2.3.13)

$$\dot{V}_b = 0; \qquad t \in [t_o, t_f] \tag{2.3.17}$$

Equations (2.3.16) and (2.3.17) imply that

$$V_b(t) = \psi(\bar{x}(t_f); t_f); \qquad t \in [t_o, t_f] \tag{2.3.18}$$

and from Equation (2.3.3),

$$V(x_o, \bar{b}+\delta b; t_o) = V(x_o, \bar{b}; t_o) + \langle V_b, \delta b\rangle + \tfrac{1}{2}\langle \delta b, V_{bb}\,\delta b\rangle \tag{2.3.19}$$

Now in order to reduce ψ to zero, we require, from Equations (2.3.18) and (2.3.19), that

$$V_b + V_{bb}\,\delta b = 0 \tag{2.3.20}$$

whence

$$\delta b = -V_{bb}^{-1}\,V_b\Big|_{t_o} = -V_{bb}^{-1}(t_o)\,\psi(\bar{x}(t_f); t_f) \tag{2.3.21}$$

The δb given by Equation (2.3.21) may now be used in Equation (2.3.14) to produce a control $u(t)$; $t \in [t_o, t_f]$ that enforces the endpoint constraint. However, because the problem is in general non-LQP, we have to limit the size of the change δb; this is done by introducing ε $(0 < \varepsilon \leqslant 1)$ into Equation (2.3.21):

$$\delta b = -\varepsilon V_{bb}^{-1}(t_o)\,\psi(\bar{x}(t_f); t_f) \tag{2.3.22}$$

If ε is sufficiently small, then Equation (2.3.14) will produce a control that: (1) reduces ψ, and (2) keeps $a(x_o, \bar{b}+\delta b; t_o) = 0$.

2.3.2. The Computational Procedure

1. Using a nominal control $\bar{u}(t)$; $t \in [t_o, t_f]$ run a nominal $\bar{x}(t)$ trajectory. Using a nominal set of multipliers $\bar{b}$, calculate the nominal cost $\bar{V}(x_o, \bar{b}; t_o)$ using Equation (2.3.2).

2. Using boundary conditions, Equation (2.3.16), integrate $\dot{a}$, $\dot{V}_x$, $\dot{V}_{xx}$ backward in time from t_f to t_o. If $|a(x_o, \bar{b}; t_o)| < \eta_1$, then go to Step 3. If not, then use the computational procedure of Section 2.2.3 to reduce $|a(x_o, \bar{b}; t_o)|$; i.e., treat the problem as a free endpoint one, by keeping $b = \bar{b}$. When $|a(x_o, \bar{b}; t_o)| < \eta_1$, go to Step 3.

3. Integrate the $\dot{V}_{xb}$ and $\dot{V}_{bb}$ equations backward along this trajectory and store $\beta_2(t)$ for $t \in [t_o, t_f]$.

4. Apply the control given by Equations (2.3.14) and (2.3.22), $\varepsilon = 1$ initially, and compute a new $x(t)$ trajectory and corresponding $\psi(x(t_f); t_f)$. If there has been an improvement in the endpoint error as measured by $|\bar{\psi}| - |\psi|$, repeat Step 2 after replacing the present nominal trajectory by the new one.† If the endpoint error is not reduced, set $\varepsilon = \varepsilon/2$ and repeat Step 4.

5. When $|a(x_o, b; t_o)| < \eta_1$ and $|\psi| < \eta_2$, stop the computation (η_1 and η_2 are chosen from numerical stability considerations).

2.3.3. Characteristics of the Algorithm

1. For the LQP problem, u, given by Equations (2.3.14) and (2.3.22), can be applied with $\varepsilon = 1$; the resulting control is the optimal control $u^o(t)$; $t \in [t_o, t_f]$. This special problem has been treated recently by Dreyfus [13].

2. The requirement that $H_{uu}^{-1}(\bar{x}, u^*, V_x; t)$ be positive-definite is not nearly as restrictive as the requirement that $H_{uu}^{-1}(\bar{x}, \bar{u}, \lambda; t)$ be positive-definite [2, 3].

3. Because we first satisfy the condition $a(x_o, \bar{b}; t_o) = 0$ and then reduce the endpoint error, we do not have to integrate the $\dot{V}_b$ equation. So the algorithm requires the integration of 2 vector differential equations fewer than the second-variation method (besides not integrating $\dot{V}_b$, we have no $\dot{h}$ equations).

2.3.4. Sufficient Conditions for an Improved Trajectory at Each Iteration

Sufficient conditions to guarantee $a(x_o, \bar{b}; t_o) < 0$ (the free endpoint problem) were given in Section 2.2.6.

For a reduction in terminal error, it is required that:

1. $V_{bb}(x_0, \bar{b}; t_0)$ be invertible‡ [Equation (2.3.22)].

Sufficient conditions for V_{bb} to be negative-definite, and hence invertible, are that:

2. $H_{uu}^{-1}(\bar{x}, u^*, V_x; t)$ be positive-definite.
3. $\psi_x^T(\bar{x}(t_f); t_f)$ has full rank s.
4. The linear system $\delta\dot{x} = (f_x + f_u\beta_1)\delta x + f_u \delta u$ be completely controllable.

Quantities in Requirement 4 are evaluated at $\bar{x}$, u^*. Note that since $a(x_0, \bar{b}; t_0)$ has been reduced to zero, $u^*(t) = \bar{u}(t)$; $t \in [t_0, t_f]$.]

† Refinements to this procedure are given in the recent paper by Gershwin and Jacobson [12].

‡ Normality.

PROOF: Condition 1 is required in order to be able to compute δb using Equation (2.3.22). We shall now show that Conditions 2-4 imply Condition 1.

The linear differential equation for V_{xb} is

$$\dot{V}_{xb} = -(f_x + f_u \beta_1)^T V_{xb}; \qquad V_{xb}(t_f) = \psi_x{}^T(x(t_f)) \tag{2.3.23}$$

The solution to this equation is

$$V_{xb}(t) = \phi^T(t_f, t) V_{xb}(t_f) \tag{2.3.24}$$

where $\phi(t, t_f)$ is the solution of

$$\dot{\phi}(t, t_f) = (f_x + f_u \beta_1) \phi(t, t_f); \qquad \phi(t_f, t_f) = I \tag{2.3.25}$$

We show that Equation (2.3.24) is the solution of Equation (2.3.23) in the following way: since ϕ is the transition matrix for the linear system $\delta \dot{x} = (f_x + f_u \beta_1) \delta x$, we have $\phi(t, t) = I = \phi(t, t_f)\phi(t_f, t)$, whence

$$\phi(t_f, t) = \phi^{-1}(t, t_f) \tag{2.3.26}$$

Substituting Equation (2.3.26) into Equation (2.3.24), we obtain

$$V_{xb}(t) = [\phi^{-1}(t, t_f)]^T V_{xb}(t_f) \tag{2.3.27}$$

So,

$$\phi^T(t, t_f) V_{xb}(t) = V_{xb}(t_f) \tag{2.3.28}$$

Differentiating Equation (2.3.28) with respect to t yields

$$\dot{\phi}^T(t, t_f) V_{xb}(t) + \phi^T(t, t_f) \dot{V}_{xb}(t) = 0$$

i.e.,

$$\dot{V}_{xb}(t) = -[\phi^T(t, t_f)]^{-1} \dot{\phi}^T(t, t_f) V_{xb}(t) \tag{2.3.29}$$

Now using Equation (2.3.25) in Equation (2.3.29),

$$\dot{V}_{xb}(t) = -(f_x + f_u \beta_1)^T V_{xb}(t) \tag{2.3.30}$$

Equation (2.3.30) is the same as Equation (2.3.23), so Equation (2.3.24) is indeed its solution.

Now,

$$-\dot{V}_{bb} = -V_{xb}^T f_u H_{uu}^{-1} f_u{}^T V_{xb} \tag{2.3.31}$$

Using Equation (2.3.24) in Equation (2.3.31),

$$-\dot{V}_{bb} = -V_{xb}^T(t_f)\,\varphi(t_f, t) f_u H_{uu}^{-1} f_u{}^T\,\phi^T(t_f, t)\,V_{xb}(t_f) \tag{2.3.32}$$

So,

$$V_{bb}(t_f) = V_{xb}^T(t_f)\,[\int_{t_o}^{t_f} \phi(t_f, t) f_u H_{uu}^{-1} f_u{}^T\,\phi^T(t_f, t)\,dt]\,V_{xb}(t_f) + V_{bb}(t_o) \tag{2.3.33}$$

Since $V_{bb}(t_f) = 0$, we have

$$V_{bb}(t_o) = -V_{xb}^T(t_f)\,[\int_{t_o}^{t_f} \phi(t_f, t) f_u H_{uu}^{-1} f_u{}^T\,\phi^T(t_f, t)\,dt]\,V_{xb}(t_f) \tag{2.3.34}$$

Now assume that the following system is completely controllable:

$$\delta\dot{x} = (f_x + f_u\beta_1)\,\delta x + f_u\,\delta u \tag{2.3.35}$$

Then $\int_{t_o}^{t_f} \phi(t_f, t) f_u f_u{}^T \phi(t_f, t)\,dt$ is positive-definite [this is a necessary and sufficient condition for the rows of $\phi(t_f, t) f_u(t)$ to be independent on the time interval $[t_o, t_f]$ [11, 14]].

Since H_{uu}^{-1} is assumed positive-definite, we see that the bracketed quantity in Equation (2.3.33) is certainly positive-definite; i.e.,

$$\int_{t_o}^{t_f} \phi(t_f, t) f_u H_{uu}^{-1} f_u{}^T\,\phi(t_f, t)\,dt \quad \text{is positive-definite} \tag{2.3.36}$$

Now we use assumption (3):

$$V_{xb}(t_f) = \psi_x{}^T \quad \text{has full rank} \quad s \tag{2.3.37}$$

Then from Equations (2.3.34), (2.3.36), and (2.3.37),

$$V_{bb}(t_o) = -\text{a positive-definite quantity}$$

i.e.,

$$V_{bb}(t_o) \quad \text{is negative-definite and is thus invertible.}$$

The fact that $V_{bb}(t_o)$ turned out to be negative-definite means that we are maximizing V with respect to the multipliers b (*cf.* Section 2.3.1).

2.3.5. Final Time t_f Given Implicitly

In some control problems the duration over which control is to be applied is not fixed *a priori*. The interval $[t_o, t_f]$ is, therefore, not specified; for a given t_o, this is equivalent to t_f being unknown. The unknown final time t_f offers us an extra degree of freedom that we may make use of by choosing t_f to minimize the cost.

We consider V to be a function of the unknown parameter t_f:

$$V(x_o, b, t_f; t_o) = \int_{t_o}^{t_f} L(x, u; t)\, dt + F(x(t_f); t_f) + \langle b, \psi(x(t_f); t_f)\rangle \qquad (2.3.38)$$

Let us assume that we do not know the optimal control $u^o(t)$; $t \in [t_o, t_f]$, the optimal multipliers b^o, or the optimal final time t_f^o. We do, however, have a nominal control $\bar{u}(t)$; $t \in [t_o, t_f]$, nominal multipliers $\bar{b}$, and a nominal final time $\bar{t}_f$.

In a similar way to that described in Sections 2.2.1 and 2.3.1, we expand $V(x, b, t_f; t)$ about $\bar{x}$, $\bar{b}$, $\bar{t}_f$ to second-order in δx, δb and δt_f:

$$\begin{aligned} V(\bar{x}+\delta x, \bar{b}+\delta b, \bar{t}_f+\delta t_f; t) &= \bar{V} + a + \langle V_x, \delta x\rangle + \langle V_b, \delta b\rangle \\ &+ V_{t_f}\delta t_f + \langle \delta x, V_{xb}\delta b\rangle + \langle V_{xt_f}, \delta x\rangle \delta t_f + \langle V_{bt_f}, \delta b\rangle \delta t_f \\ &+ \tfrac{1}{2}\langle \delta x, V_{xx}\delta x\rangle + \tfrac{1}{2}\langle \delta b, V_{bb}\delta b\rangle + \tfrac{1}{2} V_{t_f t_f}\delta t_f^{\,2} \end{aligned} \qquad (2.3.39)$$

whence†

$$V_x(\bar{x}+\delta x, \bar{b}+\delta b, \bar{t}_f+\delta t_f; t) = V_x + V_{xx}\delta x + V_{xb}\delta b + V_{xt_f}\delta t_f \qquad (2.3.40)$$

If these expressions are substituted into the DDP equation, the following differential equations are obtained:

$$-\dot{a} \quad = H - H(\bar{x}, \bar{u}, V_x; t)$$

$$-\dot{V}_x \quad = H_x + V_{xx}(f - f(\bar{x}, \bar{u}; t)) + \beta_1{}^T \overset{=0}{H_u}$$

† As in Section 2.3.1, the terms $V_{xxb}\delta x\delta b$, $V_{xbb}\delta b\delta b$, $V_{xxt_f}\delta x\delta_{t_f}$, $V_{xbt_f}\delta b\delta t_f$, $V_{xxx}\delta x\delta x$ are neglected here to avoid having to carry cumbersome terms that are neglected finally in the backward equations, for reasons given in Section 2.2.

$$
\begin{aligned}
-\dot{V}_b &= V_{xb}^T(f - f(\bar{x}, \bar{u}; t)) + \beta_2{}^T H_u^{=0} \\
-\dot{V}_{t_f} &= V_{xt_f}^T(f - f(\bar{x}, \bar{u}; t)) + \beta_3{}^T H_u^{=0} \\
-\dot{V}_{xb} &= (f_x + f_u\beta_1)^T V_{xb} \\
-\dot{V}_{xt_f} &= (f_x + f_u\beta_1)^T V_{xt_f} \qquad (2.3.41)\\
-\dot{V}_{bt_f} &= -V_{xb}^T f_u H_{uu}^{-1} f_u{}^T V_{xt_f} \\
-\dot{V}_{bb} &= -V_{xb}^T f_u H_{uu}^{-1} f_u{}^T V_{xb} \\
-\dot{V}_{xx} &= H_{xx} + f_x{}^T V_{xx} + V_{xx} f_x - (H_{ux} + f_u{}^T V_{xx})^T H_{uu}^{-1}(H_{ux} + f_u{}^T V_{xx}) \\
-\dot{V}_{t_f t_f} &= -V_{xt_f}^T f_u H_{uu}^{-1} f_u{}^T V_{xt_f}
\end{aligned}
$$

$$u(t) = u^*(t) + \beta_1(t)\delta x(t) + \beta_2(t)\delta b + \beta_3(t)\delta t_f \qquad (2.3.42)$$

where

$$
\begin{aligned}
\beta_1 &= -H_{uu}^{-1}(H_{ux} + f_u{}^T V_{xx}) \\
\beta_2 &= -H_{uu}^{-1} f_u{}^T V_{xb} \qquad (2.3.43)\\
\beta_3 &= -H_{uu}^{-1} f_u{}^T V_{xt_f}
\end{aligned}
$$

All quantities are evaluated at $\bar{x}$, b, $\bar{t}_f$, and u^* unless otherwise specified.

Let us set $b = \bar{b}$ and $t_f = \bar{t}_f$ and solve the resulting free endpoint problem† using the algorithm of Section 2.2.3. This causes $\dot{V}_b$ and $\dot{V}_{t_f}$ to be zero, and so

$$V_b(x_o, \bar{b}, \bar{t}_f; t_o) = \psi(\bar{x}(\bar{t}_f); \bar{t}_f) \qquad (2.3.44)$$

$$V_{t_f}(x_o, \bar{b}, \bar{t}_f; t_o) = V_{t_f}(\bar{x}(\bar{t}_f), \bar{b}, \bar{t}_f; \bar{t}_f) \qquad (2.3.45)$$

Now, we require

$$\psi(\bar{x}(\bar{t}_f + \delta t_f) + \delta x(\bar{t}_f + \delta t_f); \bar{t}_f + \delta t_f) = 0 \qquad (2.3.46)$$

† We assume that this problem has a solution.

and, for V to be stationary with respect to t_f,

$$V_{t_f}(x_o, \bar{b}+\delta b, \bar{t}_f+\delta t_f; t_o) = 0 \tag{2.3.47}$$

At $t = t_o$, we have, from Equation (2.3.39), that

$$V(x_o, \bar{b}+\delta b, \bar{t}_f+\delta t_f; t_o) = V + \langle V_b, \delta b\rangle + V_{t_f}\delta t_f + \langle V_{bt_f}, \delta b\rangle \delta t_f + \tfrac{1}{2}\langle \delta b, V_{bb}\delta b\rangle + \tfrac{1}{2} V_{t_f t_f}\delta t_f^2 \tag{2.3.48}$$

From Equations (2.3.44) to (2.3.48), we obtain

$$\begin{aligned} V_b + V_{bb}\delta b + V_{bt_f}\delta t_f &= 0 \\ V_{t_f} + \langle V_{bt_f}, \delta b\rangle + V_{t_f t_f}\delta t_f &= 0 \end{aligned} \tag{2.3.49}$$

i.e.,

$$\begin{bmatrix} \delta b \\ \delta t_f \end{bmatrix} = -\varepsilon \begin{bmatrix} V_{bb} & V_{bt_f} \\ V_{t_f b} & V_{t_f t_f} \end{bmatrix}^{-1} \begin{bmatrix} V_b \\ V_{t_f} \end{bmatrix} \tag{2.3.50}$$

Equation (2.3.50), together with Equations (2.3.44) and (2.3.45), allows us to solve for the δb and δt_f that reduce V_b and V_{t_f}; ε $(0 < \varepsilon \leqslant 1)$ is present to ensure that these changes are not too large.

We shall assume from now on that

$$V_{t_f t_f} \geqslant 0 \tag{2.3.51}$$

As yet we have not determined boundary conditions for Equations (2.3.41). We have that

$$V(x, b, t_f; t) = \int_t^{t_f} L(x, u; t)\,dt + F(x(t_f); t_f) + \langle b, \psi(x(t_f); t_f)\rangle \tag{2.3.52}$$

Writing Equation (2.3.52) in terms of the nominal values:

$$V(\bar{x}+\delta x, \bar{b}+\delta b, \bar{t}_f+\delta t_f; t) = \int_t^{\bar{t}_f+\delta t_f} L(x, u; t)\,dt + F(\bar{x}+\delta x+\Delta x; \bar{t}_f+\delta t_f) + \langle \bar{b}+\delta b, \psi(\bar{x}+\delta x+\Delta x; \bar{t}_f+\delta t_f)\rangle^{\dagger} \tag{2.3.53}$$

† $\bar{x}+\delta x$ is evaluated at time t.

where Δx is the change in the x trajectory over the time interval $[t, \bar{t}_f + \delta t_f]$. We require boundary conditions at $t = t_f$, so let us observe Equation (2.3.53) at $t = \bar{t}_f$.

$$V(\bar{x}+\delta x, \bar{b}+\delta b, \bar{t}_f+\delta t_f ; \bar{t}_f) = \int_{\bar{t}_f}^{\bar{t}_f+\delta t_f} L(x, u; t)\, dt + F(\bar{x}+\delta x+\Delta x; \bar{t}_f+\delta t_f) + \langle \bar{b}+\delta b, \psi(\bar{x}+\delta x+\Delta x; \bar{t}_f+\delta t_f)\rangle \qquad (2.3.54)$$

We now expand the right-hand side of Equation (2.3.54) up to second-order in δx, δb, and δt_f.

Consider the integral

$$\int_{\bar{t}_f}^{\bar{t}_f+\delta t_f} L(x, u; t)\, dt$$

$$= L(\bar{x}+\delta x, u^*+\delta u; \bar{t}_f)\,\delta t_f + \tfrac{1}{2}\frac{dL}{dt}(\bar{x}+\delta x, u^*+\delta u; \bar{t}_f)\,\delta t_f^2$$

$$= \left[L + \langle L_x, \delta x\rangle + \left\langle L_u, \frac{\partial u}{\partial x}\delta x\right\rangle + \left\langle L_u, \frac{\partial u}{\partial b}\delta b\right\rangle + \left\langle L_u, \frac{\partial u}{\partial t_f}\right\rangle \delta t_f\right]\delta t_f + \tfrac{1}{2}[L_t + \langle L_x, f\rangle + \langle L_u, \dot{u}\rangle]\,\delta t_f^2 \qquad (2.3.55)$$

All quantities are evaluated at $\bar{x}$, $\bar{b}$, $\bar{t}_f$, and u^*.

Consider the second term on the right-hand side of Equation (2.3.54):

$$F(\bar{x}+\delta x+\Delta x; \bar{t}_f+\delta t_f) = F(\bar{x}; t_f) + \langle F_x, [\delta x+\Delta x]\rangle + F_t\,\delta t_f + \langle F_{xt}, [\delta x+\Delta x]\rangle\,\delta t_f + \tfrac{1}{2}\langle \delta x+\Delta x, F_{xx}[\delta x+\Delta x]\rangle + \tfrac{1}{2}F_{tt}\,\delta t_f^2 \qquad (2.3.56)$$

where

$$\Delta x = \left[f + f_x\,\delta x + f_u\frac{\partial u}{\partial x}\delta x + f_u\frac{\partial u}{\partial b}\delta b + f_u\frac{\partial u}{\partial t_f}\delta t_f\right]\delta t_f + \tfrac{1}{2}[f_t + f_x f + f_u\dot{u}]\,\delta t_f^2 \qquad (2.3.57)$$

Quantities are evaluated at $\bar{x}$, $\bar{b}$, $\bar{t}_f$, and u^*.

Using Equation (2.3.57) in Equation (2.3.56), we obtain, to second-order, for the right-hand side of Equation (2.3.56),

$$F + \langle F_x, \delta x + f\delta t_f + f_x \delta x \delta t_f + f_u \frac{\partial u}{\partial x} \delta x \delta t_f + f_u \frac{\partial u}{\partial b} \delta b \delta t_f + f_u \frac{\partial u}{\partial t_f} \delta t_f^2$$

$$+ \tfrac{1}{2} f_t \delta t_f^2 + \tfrac{1}{2} f_x f \delta t_f^2 + \tfrac{1}{2} f_u \dot{u} \delta t_f^2 \rangle + F_t \delta t_f + \langle F_{xt}, \delta x + f\delta t_f \rangle \delta t_f$$

$$+ \tfrac{1}{2}\langle \delta x, F_{xx} \delta x \rangle + \langle \delta x, F_{xx} f \rangle \delta t_f + \tfrac{1}{2}\langle f, F_{xx} f \rangle \delta t_f^2 + \tfrac{1}{2} F_{tt} \delta t_f^2 \quad (2.3.58)$$

Consider the third term on the right-hand side of Equation (2.3.54):

$$\langle \bar{b} + \delta b, \psi(\bar{x} + \delta x + \Delta x; \bar{t}_f + \delta t_f)\rangle = \langle \bar{b} + \delta b, \psi(\bar{x}; t_f) + \psi_x[\delta x + \Delta x]$$

$$+ \psi_t \delta t_f + \psi_{xt}[\delta x + \Delta x] \delta t_f$$

$$+ \tfrac{1}{2}\psi_{xx}[\delta x + \Delta x][\delta x + \Delta x]$$

$$+ \tfrac{1}{2}\psi_{tt} \delta t_f^2 \rangle \quad (2.3.59)$$

Using Equation (2.3.57), the righ-hand side of Equation (2.3.59) becomes

$$\langle \bar{b}, \psi + \psi_x[\delta x + f\delta t_f + f_x \delta x \delta t_f + f_u \frac{\partial u}{\partial x} \delta x \delta t_f + f_u \frac{\partial u}{\partial b} \delta b \delta t_f + f_u \frac{\partial u}{\partial t_f} \delta t_f^2$$

$$+ \tfrac{1}{2} f_t \delta t_f^2 + \tfrac{1}{2} f_x f \delta t_f^2 + \tfrac{1}{2} f_u \dot{u} \delta t_f^2 + \psi_t \delta t_f + \psi_{xt}[\delta x + f\delta t_f] \delta t_f$$

$$+ \tfrac{1}{2}\psi_{xx} \delta x \delta x + \psi_{xx} f \delta x \delta t_f + \tfrac{1}{2}\psi_{xx} ff \delta t_f^2 + \tfrac{1}{2}\psi_{tt} \delta t_f^2 \rangle$$

$$+ \langle \delta b, \psi + \psi_x[\delta x + f\delta t_f] + \psi_t \delta t_f \rangle \quad (2.3.60)$$

Adding Expressions (2.3.55), (2.3.58), and (2.3.60),

$$F + \langle \bar{b}, \psi \rangle + [L + \langle F_x + \psi_x^T \bar{b}, f \rangle + F_t + \langle \bar{b}, \psi_t \rangle] \delta t_f + \langle F_x + \psi_x^T \bar{b}, \delta x \rangle$$

$$+ \langle \psi, \delta b \rangle + \langle L_x + f_x^T(F_x + \psi_x^T \bar{b}) + \frac{\partial u^T}{\partial x}[L_u + f_u^T(F_x + \psi_x^T \bar{b})]$$

$$+ F_{xt} + \psi_{xt}^T \bar{b} + (F_{xx} + \bar{b}\psi_{xx})f, \delta x\rangle \delta t_f + \langle \delta b, \psi_x \delta x\rangle$$

$$+ \langle \delta b, \psi_t + \psi_x f\rangle \delta t_f + \langle \frac{\partial u^T}{\partial b}[L_u + f_u{}^T(F_x + \psi_x{}^T \bar{b})], \delta b\rangle \delta t_f$$

$$+ \tfrac{1}{2}[L_t + \langle F_x + \psi_x{}^T \bar{b}), f_t\rangle + \langle L_x + f_x{}^T(F_x + \psi_x{}^T \bar{b}), f\rangle$$

$$+ \langle L_u + f_u{}^T(F_x + \psi_x{}^T \bar{b}), \dot{u}\rangle + 2\langle F_{xt} + \psi_{xt}^T \bar{b}, f\rangle + F_{tt} + \langle \bar{b}, \psi_{tt}\rangle$$

$$+ \langle f, (F_{xx} + \bar{b}\psi_{xx})\rangle] \delta t_f{}^2 + 2\langle L_u + f_u{}^T(F_x + \psi_x{}^T \bar{b}), \frac{\partial u}{\partial t_f} \delta t_f{}^2\rangle$$

$$+ \tfrac{1}{2}\langle \delta x, (F_{xx} + \bar{b}\psi_{xx})\delta x\rangle \tag{2.3.61}$$

The left-hand side of Equation (2.3.54), expanded to second-order, is

$$V(\bar{x}, \bar{b}, \bar{t}_f; t_f) + \langle V_x, \delta x\rangle + \langle V_b, \delta b\rangle + V_{t_f}\delta t_f + \langle V_{xt_f}, \delta x\rangle \delta t_f$$
$$+ \langle \delta x, V_{xb}\delta b\rangle + \langle \delta b, V_{bt_f}\rangle \delta t_f + \tfrac{1}{2}\langle \delta x, V_{xx}\delta x\rangle$$
$$+ \tfrac{1}{2}\langle \delta b, V_{bb}\delta b\rangle + \tfrac{1}{2} V_{t_f t_f}\delta t_f{}^2 \tag{2.3.62}$$

Equating coefficients of like powers of $\delta x, \delta b$, and δt_f between Equations (2.3.61) and (2.3.62) yields

$$\begin{aligned}
V_x &= F_x + \psi_x{}^T \bar{b} \\
V_b &= \psi \\
V_{t_f} &= H + F_t + \langle \bar{b}, \psi_t\rangle \\
V_{xb} &= \psi_x{}^T \\
V_{xx} &= F_{xx} + \bar{b}\psi_{xx} \\
V_{bb} &= 0 \\
V_{xt_f} &= F_{xt} + \psi_{xt}^T \bar{b} + H_x + V_{xx} f + (\partial u^T/\partial x)\, \overset{=0}{H_u} \\
V_{bt_f} &= \psi_t + \psi_x f + (\partial u^T/\partial x) \overset{=0}{H_u} \\
V_{t_f t_f} &= H_t + F_{tt} + \langle \bar{b}, \psi_{tt}\rangle + \langle H_x, f\rangle + \langle \overset{=0}{H_u}, \dot{u}\rangle + 2\langle \overset{=0}{H_u}, \partial u/\partial t_f\rangle \\
&\quad + 2\langle F_{xt} + \psi_{xt}^T b, f\rangle + \langle f, V_{xx} f\rangle
\end{aligned} \tag{2.3.63}$$

where

$$H = L + \langle V_x, f \rangle \tag{2.3.64}$$

All quantities evaluated at $\bar{x}$, $\bar{b}$, $\bar{t}_f$, and u^*.

Equations (2.3.63) provide boundary conditions for the differential Equations (2.3.41). Though the derivation of these boundary conditions is rather tedious, we feel that this dynamic programming approach is more direct and easier to follow than the second-variation transversality derivations described by Mitter [2] and McReynolds and Bryson [3].†

2.3.6. The Computational Procedure

The procedure is much the same as that of Section 2.3.2; however, now we have an extra parameter t_f additional to b. The extra differential equations for V_{xt_f}, V_{bt_f} and $V_{t_f t_f}$ must, therefore, be integrated. At $t = t_o$, changes δb and δt_f are computed using Equation (2.3.50), and $u(t)$ on the forward run is given by Equation (2.3.42).

2.3.7. Characteristics of the Algorithm

1. The algorithm requires the integration of 3 sets of differential equations less than the second-variation methods (there are no $\dot{h}$ equations, and $\dot{V}_b$ and $\dot{V}_{t_f}$ are zero, because of the way in which the algorithm is implemented).

2. The requirement that $V_{t_f t_f} > 0$ may be quite restrictive because, even with an LQP problem with linear endpoint constraints, it is unlikely that $V_{t_f t_f} > 0$ will be satisfied for any $\bar{x}$, $\bar{b}$, or $\bar{t}_f$.

2.3.8. Sufficient Conditions for an Improved Trajectory at Each Iteration

In addition to the conditions of Section 2.3.4, it is required that

$$V_{t_f t_f}(\bar{x}, \bar{b}, t_f; t_o) > 0 \tag{2.3.65}$$

This is a sufficient condition to guarantee that the matrix in Equation (2.3.49) is invertible.

PROOF: We have shown that under certain conditions, given in Section 2.3.4, $V_{bb}(t_o)$ is negative-definite.

† The boundary conditions given by Mitter [2] and McReynolds and Bryson [3] do not agree with ours, and, therefore, appear to be in error. We thank Mukund Desai for pointing out an error in an earlier derivation of our results.

The matrix partition lemma states that

$$\begin{bmatrix} V_{bb} & V_{bt_f} \\ V_{t_f b} & V_{t_f t_f} \end{bmatrix}^{-1} = \begin{bmatrix} B_{11} & B_{12} \\ B_{21} & B_{22} \end{bmatrix} \tag{2.3.66}$$

where

$$\begin{aligned} B_{11} &= V_{bb}^{-1} + V_{bb}^{-1} V_{bt_f} z^{-1} V_{t_f b} V_{bb}^{-1} \\ B_{12} &= -V_{bb}^{-1} V_{bt_f} z^{-1} \\ B_{21} &= -z^{-1} V_{t_f b} V_{bb}^{-1} \\ B_{22} &= z^{-1} \\ z &= V_{t_f t_f} - V_{bt_f}^T V_{bb}^{-1} V_{bt_f} \end{aligned} \tag{2.3.67}$$

Since V_{bb} is invertible, we require only that z be invertible. Since V_{bb} is negative-definite, the scalar $-V_{bt_f} V_{bb}^{-1} V_{bt_f} \geqslant 0$; so if $V_{t_f t_f}$ is positive, z is positive and hence invertible.

2.3.9. Computed Example

In this section we solve the Dreyfus rocket problem [15] using the second-order algorithm of Section 2.3.2.

Dreyfus [15] used a first-order method to solve a rocket problem; Mitter [2] attempted the same problem using his second-variation method. The problem is to launch a rocket in fixed time to a given altitude and with a given final vertical velocity component and maximum final horizontal velocity component.

We shall assume that the motion of the rocket is governed by the following differential equations:†

$$\begin{aligned} \dot{x}_1 &= x_3; & x_1(t_0) &= 0 \\ \dot{x}_2 &= x_4; & x_2(t_o) &= 0 \\ \dot{x}_3 &= a \cos u; & x_3(t_o) &= 0 \\ \dot{x}_4 &= a \sin u - g; & x_4(t_o) &= 0 \end{aligned} \tag{2.3.68}$$

† The approximations and assumptions necessary to arrive at these equations are given by Dreyfus [15].

where x_1 is the horizontal range (feet), x_2 is the altitude (feet), x_3 is the horizontal velocity component (feet per second), x_4 is the vertical velocity component (feet per second), and u is the inclination of the thrust to the horizontal (radians). We are required to choose $u(t)$; $t \in [t_o, t_f]$ to maximize $x_3(t_f)$, subject to the constraints that

$$\begin{aligned} x_2(t_f) &= 100{,}000 \text{ ft} \\ x_4(t_f) &= 0 \text{ ft/sec} \end{aligned} \tag{2.3.69}$$

The initial time t_o is given as zero, and t_f is 100 sec; x_1 does not enter the control problem and so, need not be considered. In order to change the problem to a minimization, we minimize

$$V = -x_3(t_f) = \int_0^{t_f} -a \cos u \, dt \tag{2.3.70}$$

The relevant differential equations are

$$\begin{aligned} \dot{x}_2 &= x_4 \\ \dot{x}_4 &= a \sin u - g \end{aligned} \tag{2.3.71}$$

a is given as 64 ft/sec^2 and g as 32 ft/sec^2. We adjoin the constraints, Equation (2.3.69), to the cost function, Equation (2.3.70), using Lagrange multipliers b_1 and b_2:

$$V(x_o, b; t_o) = \int_0^{100} -a \cos u \, dt + b_1(x_2 - 100{,}000) + b_2 x_4 \tag{2.3.72}$$

For this problem,

$$\begin{aligned} H &= -a \cos u + V_{x_2} x_4 + V_{x_4}(a \sin u - g) \\ H_{xx} &= 0 \\ H_{ux} &= 0 \end{aligned} \tag{2.3.73}$$

$$\begin{aligned} H_u &= a \sin u + V_{x_4} a \cos u \\ H_{uu} &= a \cos u - V_{x_4} a \sin u \end{aligned} \tag{2.3.74}$$

From Equation (2.3.74),

$$u^* = \operatorname{atan}(-V_{x_4}) \tag{2.3.75}$$

whence

$$H_{uu}(\bar{x}, u^*, V_x; t) = a/\cos u^* \tag{2.3.76}$$

From Equation (2.3.76), it is clear that H_{uu} is positive for u^* between $-(\pi/2)$ and $\pi/2$, regardless of the nominal trajectory.
From Equation (2.3.72),

$$V_x(t_f) = \begin{bmatrix} b_1 \\ \\ b_2 \end{bmatrix}$$

$$V_b(t_f) = \begin{bmatrix} x_2 - 100{,}000 \\ \\ x_4 \end{bmatrix} \tag{2.3.77}$$

$$V_{xb}(t_f) = I, \qquad V_{xx}(t_f) = 0, \qquad V_{bb}(t_f) = 0$$

Consider the free endpoint problem, Equation (2.3.72), with $b = \bar{b}$; we have, since $H_{xx} = H_{ux} = V_{xx}(t_f) = 0$, that

$$V_{xx}(t) = 0; \qquad t \in [t_o, t_f] \tag{2.3.78}$$

The relevant backward equations are

$$\begin{aligned} -\dot{a} &= H - H(\bar{x}, \bar{u}, V_x; t) \\ -\dot{V}_x &= H_x \end{aligned} \tag{2.3.79}$$

and, from Equation (2.3.14),

$$u(t) = u^*(t) \qquad [\text{since} \quad \beta_1(t) = 0] \tag{2.3.80}$$

In view of Equations (2.3.79) and (2.3.80), it is clear that only one iteration is required to find the optimal control for the free endpoint problem with $b = \bar{b}$.

Having solved this problem, we integrate

$$\begin{aligned} -\dot{V}_{xb} &= f_x^{\,T} V_{xb} \\ -\dot{V}_{bb} &= -V_{xb}^T f_u H_{uu}^{-1} f_u^{\,T} V_{xb} \end{aligned} \tag{2.3.81}$$

using the boundary conditions given by Equations (2.3.77).
Starting with nominal multipliers,

$$\bar{b}_1 = 0.1 \qquad \bar{b}_2 = 1.0 \tag{2.3.82}$$

and the same nominal control used by Mitter [2] and Dreyfus [15]:

$$\bar{u}(t) = 1.6 - 1.5(t/100) \tag{2.3.83}$$

the procedure of Section 2.3.2 was applied, and the optimal trajectory was obtained in 6 iterations (1 minute of computing time on the IBM 7090).†

The progress of the algorithm is illustrated in Figures 2.4-2.7. Figure 2.4 shows the cost V, Equation (2.3.72), as a function of iteration number. Notice the initial fall in the cost at the first iteration; this is due to the minimization of V with $b_1 = \bar{b}_1$ and $b_2 = \bar{b}_2$. Thereafter, the cost rises as b_1 and b_2 are changed.

Figure 2.5 shows the control functions for various iterations. Figures 2.6 and 2.7 are plots of $x_2(t_f)$ and $x_4(t_f)$ versus iteration number.

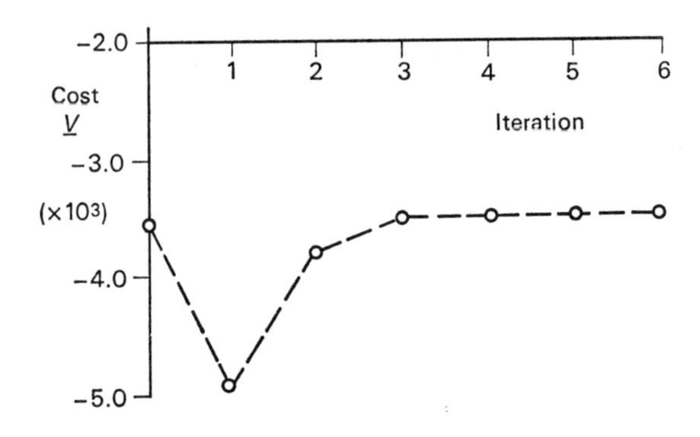

Figure 2.4. Dreyfus rocket problem; cost versus iteration number.

† A fourth-order Runge-Kutta integration routine was used for all the integrations; 300 integration steps were used.

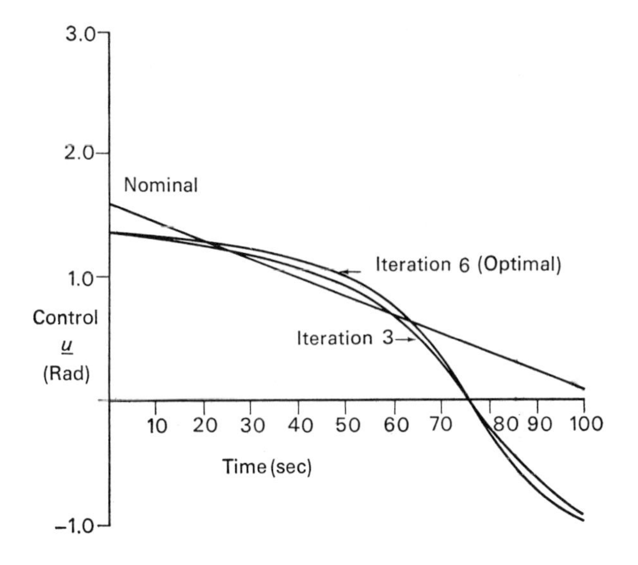

Figure 2.5. Dreyfus rocket problem: control functions.

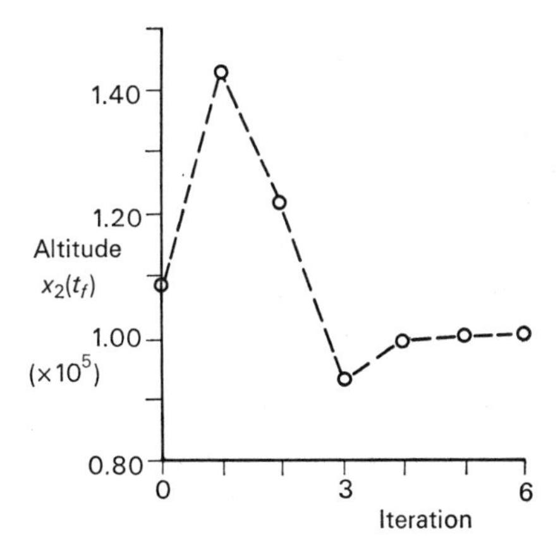

Figure 2.6. Dreyfus rocket problem: Final altitude versus iteration number.

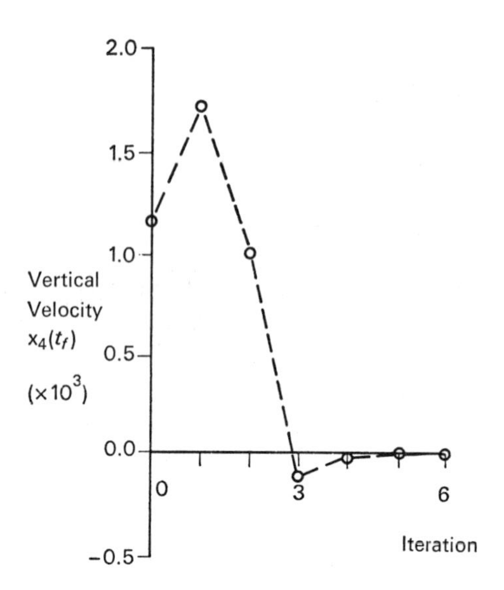

Figure 2.7. Dreyfus rocket problem: final vertical velocity versus iteration number.

At the sixth iteration, x_2 is within 0.1 ft of 100,000 ft and x_4 is within 0.1 ft/sec of zero. $-V = x_3 = 3508.0$ ft/sec and $b_1{}^o = 0.0632$; $b_2{}^o = 1.4900$. The analytic solution for this problem is known to be $x_3 = 3507.9$ ft/sec.

Dreyfus obtained results of similar accuracy but required 29 iterations of a first-order method that used 3 minutes of IBM 7090 time.

Mitter used a penalty function approach when attempting to solve this problem using the second-variation algorithm. Mitter found that convergence was hampered by $H_{uu}^{-1}(\bar{x}, \bar{u}, V_x; t)$ not being positive-definite along the nominal trajectories; in addition, great difficulty was encountered in satisfying the end constraints.

The second-order algorithm of Section 2.3.2 had little difficulty in solving this problem. Adjoining the end constraints to the cost function by means of the Lagrange multipliers b_1 and b_2 actually simplified the problem, since V_{xx} turned out to be zero on the interval $[t_o, t_f]$.

2.4. A NEW SECOND-ORDER ALGORITHM FOR FREE ENDPOINT PROBLEMS WITH CONTROL INEQUALITY CONSTRAINTS

2.4.1. The Derivation

We shall consider a class of control problems where control constraints of the following form are present:

$$g(u; t) \leqslant 0 \tag{2.4.1}$$

where g is a $p \leqslant m$-vector function.

In the past it has not been obvious how to incorporate this type of constraint into second-variation and second-order algorithms. Noton *et al.* [16] have stated that any iterative procedure based on expansions appears to be inapplicable to control inequality constrained problems. We shall show, however, that control inequality constraints can be included in our second-order analyses.

We shall treat the following problem:

$$\dot{x} = f(x, u; t); \qquad x(t_o) = x_o \tag{2.4.2}$$

$$V(x_o; t_o) = \int_{t_o}^{t_f} L(x, u; t)\, dt + F(x(t_f)) \tag{2.4.3}$$

$$g(u; t) \leqslant 0 \tag{2.4.4}$$

There are no endpoint constraints, and t_f is given explicitly.

We make the following assumption: the optimal control function $u^o(t)$ is continuous on the whole interval $[t_o, t_f]$; i.e., if and when a control hits or leaves a constraint, it does so without a sudden jump.† Figure 2.8 illustrates times t_a and t_i at which the constraint becomes active and inactive, respectively.

Referring to Equation (1.4.13), we have: ‡

$$-\frac{\partial \bar{V}}{\partial t} - \frac{\partial a}{\partial t} - \left\langle \frac{\partial V_x}{\partial t}, \delta x \right\rangle - \tfrac{1}{2}\left\langle \delta x, \frac{\partial V_{xx}}{\partial t}\, \delta x \right\rangle$$

$$= \min_{\delta u} [L(\bar{x}+\delta x, \bar{u}+\delta u; t) + \langle V_x + V_{xx}\delta x, f(\bar{x}+\delta x, \bar{u}+\delta u; t)\rangle] \tag{2.4.5}$$

The right-hand side of Equation (2.4.5) is

$$\min_{\delta u} [H(\bar{x}+\delta x, \bar{u}+\delta u, V_x; t) + \langle V_{xx}\delta x, f(\bar{x}+\delta x, \bar{u}+\delta u; t)\rangle] \tag{2.4.6}$$

---

† This assumption is not overly restrictive; many control inequality constrained problems exhibit this behavior. An exception is the bang-bang problem where $H_{uu} = 0$ and the constraint g is of the form $|u| \leqslant 1$;solutions to this problem are obtained in Chapter 3.

‡ Again, $V_{xxx}\delta x \delta x$ is neglected for reasons given in Section 2.2.

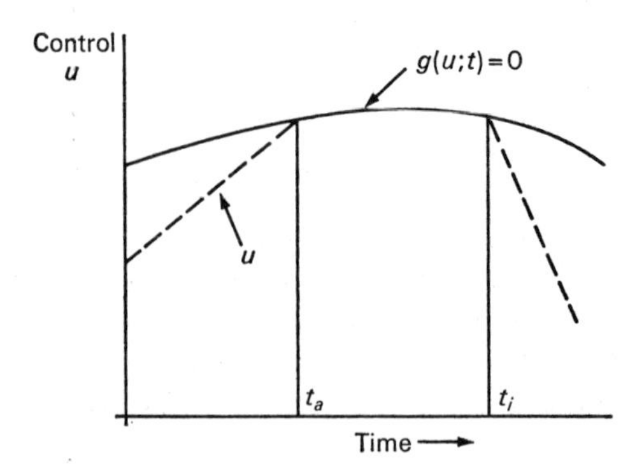

Figure 2.8. Illustration of a control inequality constraint.

As in Section 2.2.1, let us consider the case where $\delta x = 0$; we must now carry out the minimization subject to constraint (2.4.4). Define the set

$$U = \{u: g(u; t) \leqslant 0\} \tag{2.4.7}$$

Expression (2.4.6) becomes, for $\delta x = 0$,

$$\min_{\delta u} [H(\bar{x}, \bar{u} + \delta u, V_x; t)], \qquad \bar{u} + \delta u \in U \tag{2.4.8}$$

Let $\hat{u}$ be the control that minimizes H with respect to u, subject to $u \varepsilon U$.
Expression (2.4.8) becomes

$$H(\bar{x}, \hat{u}, V_x; t) \tag{2.4.9}$$

(The constrained minimization in Expression (2.4.8) is a nonlinear programming problem. In control problems it often happens that there are only a few controls and a few simple constraints on u; the minimization is then performed easily).

If the minimization is performed and it turns out that no constraints are active (i.e., strict inequality holds in Equation (2.4.4), for $u = \hat{u}$; so $\hat{u} = u^*$), then the algorithm is the same as that of Section 2.2.

Assume that $\hat{p}$ of the constraints are active, $0 < \hat{p} \leqslant p$, and refer to them as $\hat{g}(u; t)$:

$$\hat{g}(\hat{u}; t) = 0 \tag{2.4.10}$$

Adjoining Equation (2.4.10) to (2.4.9), using a vector Lagrange multiplier λ of dimension $\hat{p}$, we obtain:†

$$J(\bar{x}, \hat{u}, \lambda, V_x; t) = H(\bar{x}, \hat{u}, V_x; t) + \langle \lambda, \hat{g}(\hat{u}; t) \rangle \tag{2.4.11}$$

Under certain assumptions, given in Section 2.4.4, the following equations are necessary for determining λ and $\hat{u}$:

$$\partial J / \partial u = H_u(\bar{x}, \hat{u}, V_x; t) + \hat{g}_u^T(\hat{u}; t)\lambda = 0 \tag{2.4.12}$$

$$\partial J / \partial \lambda = \hat{g}(\hat{u}; t) = 0 \tag{2.4.13}$$

Assume now that variations δx in x are introduced at time t and that all the constraints $\hat{g}(\hat{u} + \delta u; t)$ remain active.‡ This will be true at all times t except those at which a constraint is just becoming active or inactive; ignore such boundary points as t_a and t_i for the moment.

Reintroducing δx into Equation (2.4.11), we obtain the following expression:

$$\min_{\delta u}[H(\bar{x} + \delta x, \hat{u} + \delta u, V_x; t) + \langle V_{xx}\delta x, f(\bar{x} + \delta x, \hat{u} + \delta u; t) \rangle$$
$$+ \langle \lambda + \delta\lambda, \hat{g}(\hat{u} + \delta u; t) \rangle] \tag{2.4.14}$$

$\delta\lambda$ is also present on the reintroduction of δx, since the constraints $\hat{g}$ at $\hat{u} + \delta u$ are assumed to remain active, and so, λ must change to $\lambda + \delta\lambda$ to ensure this.

Analogous conditions to Equations (2.4.12) and (2.4.13) are

$$H_u(\bar{x} + \delta x, \hat{u} + \delta u, V_x; t) + f_u^T(\bar{x} + \delta x, \hat{u} + \delta u; t) V_{xx} \delta x$$
$$+ \hat{g}_u^T(\hat{u} + \delta u; t)(\lambda + \delta\lambda) = 0 \tag{2.4.15}$$

and

$$\hat{g}(\hat{u} + \delta u; t) = 0 \tag{2.4.16}$$

Expanding to first-order§ about $\bar{x}$, $\hat{u}$ and using Equations (2.4.12) and (2.4.13), we obtain

$$(H_{uu} + \lambda\hat{g}_{uu})\delta u + \hat{g}_u^T \delta\lambda = -(H_{ux} + f_u^T V_{xx})\delta x \tag{2.4.17}$$

† At different times $t \in [t_0, t_f]$, $\hat{p}$ may have a different value and $\hat{g}(\hat{u}; t)$ may be different.

‡ That is, assume that the $\hat{g}$ are active; $\lambda_i > 0$, $(i = 1, \ldots, \hat{p})$.

§ Reasons for expanding to first-order only are similar to those given in Section 2.2.1. Here, $\lambda\hat{g}_{uu} \triangleq \sum_i \lambda_i \hat{g}_{i_{uu}}$.

$$\hat{g}_u \delta u = 0 \tag{2.4.18}$$

All quantities are evaluated at $\bar{x}$, $\hat{u}$; t.

From these equations,

$$\delta\lambda = -[\hat{g}_u(H_{uu}+\lambda\hat{g}_{uu})^{-1}\hat{g}_u{}^T]^{-1}\hat{g}_u(H_{uu}+\lambda\hat{g}_{uu})^{-1}(H_{ux}+f_u{}^T V_{xx})\delta x \tag{2.4.19}$$

and

$$\delta u = -(H_{uu}+\lambda\hat{g}_{uu})^{-1}\{I_m - \hat{g}_u{}^T[\hat{g}_u(H_{uu}+\lambda\hat{g}_{uu})^{-1}\hat{g}_u{}^T]^{-1} \cdot \hat{g}_u(H_{uu}+\lambda\hat{g}_{uu})^{-1}\}(H_{ux}+f_u{}^T V_{xx})\delta x \tag{2.4.20}$$

where I_m is the unit matrix of dimension m (the number of controls).

Expanding Equation (2.4.14) to second-order, substituting Expressions (2.4.19) and (2.4.20) for $\delta\lambda$ and δu, and using Equations (2.4.12) and (2.4.13), the following expression is obtained:

$$H + \langle H_x + V_{xx} f, \delta x\rangle + \tfrac{1}{2}\langle \delta x, [H_{xx} + f_x{}^T V_{xx} + V_{xx} f_x - (H_{ux}+f_u{}^T V_{xx})^T Z^T (H_{uu}+\lambda\hat{g}_{uu})^{-1} Z (H_{ux}+f_u{}^T V_{xx})]\delta x\rangle \tag{2.4.21}$$

where

$$Z = I_m - \hat{g}_u{}^T[\hat{g}_u(H_{uu}+\lambda\hat{g}_{uu})^{-1}\hat{g}_u{}^T]^{-1}\hat{g}_u(H_{uu}+\lambda\hat{g}_{uu})^{-1} \tag{2.4.22}$$

Equation Expression (2.4.21) to the left hand side of Equation (2.4.5), the following equations are obtained in the manner described in Section 2.2.1:

$$-\dot{a} = H - H(\bar{x}, \bar{u}, V_x; t)$$

$$-\dot{V}_x = H_x + V_{xx}(f - f(\bar{x}, \bar{u}; t))$$

$$-\dot{V}_{xx} = H_{xx} + f_x{}^T V_{xx} + V_{xx} f_x - (H_{ux} + f_u{}^T V_{xx})^T Z^T (H_{uu}+\lambda\hat{g}_{uu})^{-1} Z (H_{ux}+f_u{}^T V_{xx}) \tag{2.4.23}$$

$$u = \hat{u} + \hat{\beta}\,\delta x \tag{2.4.24}$$

where

$$\hat{\beta} = -(H_{uu}+\lambda\hat{g}_{uu})^{-1} Z (H_{ux}+f_u{}^T V_{xx}) \tag{2.4.25}$$

and $\hat{u}$ is chosen by $\min_{u \in U} H(\bar{x}, u, V_x; t)$ which yields also the $\hat{g}(\hat{u}; t)$; λ is given by Equations (2.4.12) and (2.4.13). Unless otherwise stated, all quantities are evaluated at $\bar{x}, \hat{u}; t$. Boundary conditions are the same as before, namely Equations (2.2.22) to (2.2.24).

At times when no constraints are active, the above equations reduce to those of Section 2.2.1 (i.e., $Z = I_m$).

In the above derivations, $H_{uu} + \lambda \hat{g}_{uu}$ has been assumed positive-definite. Since $H_{uu} + \lambda \hat{g}_{uu}$ is evaluated at $\hat{u}$ and not the nominal $\bar{u}$, global, strict convexity with respect to u of $H + \langle \lambda, g \rangle$ is not required. Local, strict convexity at $\hat{u}$ is sufficient; many problems exhibit this property.

If there is only one control, then, from Equation (2.4.18), $\delta u = 0$ if $\hat{g}_u \neq 0$ and $Z = \hat{\beta} = 0$ even if $H_{uu} + \lambda \hat{g}_{uu}$ is not positive. Note that in this case the Riccati equation degenerates into a linear matrix equation.

At boundary points where a constraint ceases to be active or inactive, Z will change discontinuously; however, $\hat{u}$ is continuous. It follows, then, that only $\dot{V}_{xx}$ suffers a discontinuity.

From Equation (2.4.25), note that when running forward and generating the new trial trajectory, $u(t)$ will be discontinuous at times of boundary points† of $\hat{g}$ because of the presence of the discontinuity in Z. However, this discontinuity does not affect the cost to second-order; this is intuitively so, and is proved in Section 3.3.2. The discontinuity in the forward $u(t)$ can be overcome easily, if desired, by using the computational trick of Section 2.2.5. Since V_x and V_{xx} are continuous, the $u(t)$ so generated will be continuous (on an optimal trajectory, $u = \hat{u} = u^o$, which is continuous).

2.4.2. The Computational Procedure

The computational procedure for this algorithm is the same as that described in Section 2.2.3, except that the minimization of H with respect to u is performed for $u \in U$; this produces $\hat{u}$ and $\hat{g}$. Using Equations (2.4.12) and (2.4.13), λ is then calculated, which enables Z and $\hat{\beta}$ to be calculated using Equations (2.4.22) and (2.4.25). The nominal control $\bar{u}(t)$ is assumed to satisfy $g(\bar{u}; t) \leqslant 0$.

2.4.3. Characteristics of the Algorithm

1. Control inequality constraints of the type $g(u; t) \leqslant 0$ can be handled, provided the optimal control function is continuous. It is believed that the algorithm is the only second-order method available that can treat these problems.

† In the neighborhood of boundary points, $u(t)$ may violate Inequality (2.4.4). The computational trick of Section 2.2.5 overcomes this difficulty.

2. The procedure does not exhibit one step onvergence for the LQP problem with linear control constraints because the optimal cost $V^{o}(x; t)$, for that problem, is not quadratic.

3. The requirement that $H_{uu}(\bar{x}, \hat{u}, V_x; t) + \lambda \hat{g}_{uu}(\hat{u}; t)$ be positive-definite is rather restrictive. The control problem where this matrix is identically zero will be treated in Chapter 3.

2.4.4. Sufficient Conditions for a Reduction in Cost at Each Iteration

In this and the next section it should be remembered that $\hat{g}(\hat{u}; t) = 0$ and $\bar{u} \in U$. Sufficient conditions to guarantee $a(\bar{x}; t_1) < 0$, and hence a reduction in cost for δx sufficiently small, are that, for $t \in [t_o, t_f]$,

(1) $H(\bar{x}, \hat{u}, V_x; t) < H(\bar{x}, \bar{u}, V_x; t); \quad \hat{u} \neq \bar{u}$.

(2) The solutions of the differential equations be bounded. Sufficient conditions to allow the calculation of λ, $\hat{\beta}$ and Z, are that, for $t \in [t_0, t_f]$,

(3) $\hat{g}_u(\hat{u}; t)$ have full rank $\hat{p}$ and $[\hat{g}_u(\hat{u}; t) \vdots H_u(\bar{x}, \hat{u}, V_x; t)]$ have rank $\hat{p}$.

(4) $[H_{uu}(\bar{x}, \hat{u}, V_x; t) + \lambda \hat{g}_{uu}(\hat{u}; t)]^{-1}$ be positive-definite.

PROOF:

$$a(\bar{x}; t_1) = \int_{t_f}^{t_1} [H(\bar{x}, \hat{u}, V_x; t) - H(\bar{x}, \bar{u}, V_x; t)]\, dt \tag{2.4.26}$$

A sufficient condition for $a(\bar{x}; t_1) < 0$ is clearly

$$H(\bar{x}, \hat{u}, V_x; t) < H(\bar{x}, \bar{u}, V_x; t); \qquad \hat{u} \neq \bar{u},\ \hat{u} \text{ and } \bar{u} \in U \tag{2.4.27}$$

Clearly, the differential equations are required to have bounded solutions. From linear equation theory [17, 18], necessary and sufficient conditions for λ to be determined from Equations (2.4.12) and (2.4.13) are that:

$$\hat{g}_u^T(\hat{u}; t) \qquad \text{have full rank } \hat{p} \tag{2.4.28}$$

$$[\hat{g}_u^T(\hat{u}; t) \vdots H_u(\bar{x}, \hat{u}, V_x; t)] \qquad \text{have rank } \hat{p} \tag{2.4.29}$$

The positive-definiteness of

$$H_{uu}(\bar{x}, \hat{u}, V_x; t) + \lambda \hat{g}_{uu}(\hat{u}; t)$$

allows the calculation of $\hat{\beta}$ and, together with $\hat{g}_u^T(\hat{u}; t)$ having rank $\hat{p}$, allows the calculation of Z.

2.4.5. Control Inequality Constraints: A New First-Order Algorithm

As in Section 2.2.7, a first-order algorithm emerges as a special case of the second-order one:

$$-\dot{a} = H - H(\bar{x}, \bar{u}, V_x; t); \qquad a(t_f) = 0 \tag{2.4.30}$$

$$-\dot{V}_x = H_x \qquad ; \qquad V_x(t_f) = F_x(\bar{x}(t_f); t_f) \tag{2.4.31}$$

All quantities are evaluated at $\bar{x}, \hat{u}; t$ unless stated otherwise.

The new control that is applied is

$$u(t) = \hat{u}(t); \qquad t \in [t_1, t_f] \tag{2.4.32}$$

2.4.6. Characteristics of the Algorithm

The implementation and characteristics of the algorithm are similar to those of Sections 2.2.7 and 2.2.8.

2.4.7. Computed Examples

Consider again the Rayleigh Equation (2.2.64). The following control constraints are introduced:

$$|u| \leqslant 1 \tag{2.4.33}$$

Note that $H_{uu} = 2$ and the constraints $g(u; t)$ are of the form

$$u - 1 \leqslant 0 \qquad \text{if} \quad u > 0 \tag{2.4.34}$$

$$u + 1 \geqslant 0 \qquad \text{if} \quad u < 0 \tag{2.4.35}$$

It is clear, therefore, that $H_{uu} + \lambda g_{uu}$ is positive for all u. It is easily seen that when the constraint is inactive, $Z = 1$ and when it is active, $Z = 0$ and the Riccati equation becomes a linear one.

Starting from the same nominal trajectory as before, the new second-order algorithm found the optimal solution in 3 iterations. Figure 2.9 shows the cost as a function of iteration number and Figure 2.10 shows the control function for various iterations. Note the jumps in the control along non-optimal trajectories and observe that they disappear when the optimal trajectory is reached.

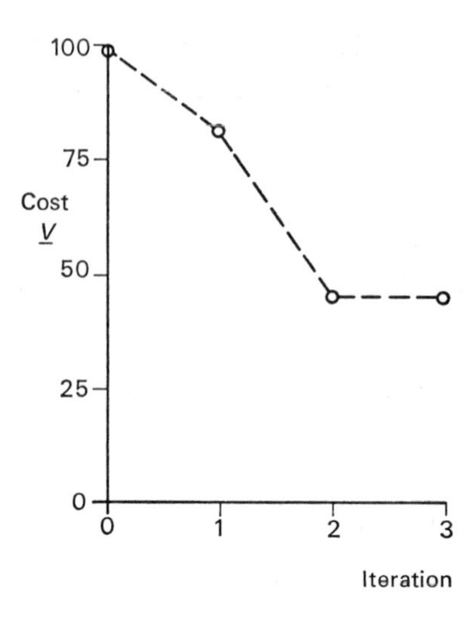

Figure 2.9. Constrained rayleigh problem: cost versus iteration number

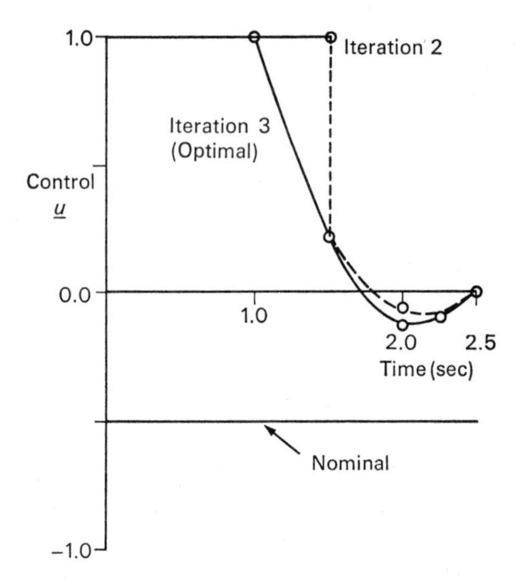

Figure 2.10. Constrained rayleigh problem: control functions.

It should be noted that the Riccati equation now has a bounded solution on the whole interval $[t_o, t_f]$, even along nonoptimal trajectories. This is because, when $Z = \hat{\beta} = 0$ along the constraint, it becomes a linear equation that cannot have an unbounded solution. The time interval over which $Z=1$ and $\hat{\beta}\neq 0$ is too small for the Riccati equation to produce an unbounded solution.

2.5. A NEW SECOND-ORDER ALGORITHM FOR FIXED ENDPOINT PROBLEMS WITH CONTROL INEQUALITY CONSTRAINTS

2.5.1. The Derivation

We consider now the control inequality constrained problem with the additional endpoint equality constraints:

$$\psi(x(t_f))=0 \tag{2.5.1}$$

We assume that t_f is given explicitly; the case where t_f is given implicitly is straightforward, but tedious.

As in Section 2.3.1, we adjoin the s-dimensional vector of constraints ψ to the cost V using an s-dimensional vector Lagrange multiplier b:

$$V(x_o, b; t_o) = \int_{t_o}^{t_f} L(x, u; t)\,dt + F(x(t_f)) + \langle b, \psi(x(t_f))\rangle \tag{2.5.2}$$

In a similar way to that described in Section 2.3.1, the following results may be obtained:

$$\begin{aligned}
-\dot{a} &= H - H(\bar{x}, \bar{u}, V_x; t)\\
-\dot{V}_x &= H_x + V_{xx}(f - f(\bar{x}, \bar{u}; t))\\
-\dot{V}_b &= V_{xb}^T(f - f(\bar{x}, \bar{u}; t))\\
-\dot{V}_{xb} &= (f_x^T + \hat{\beta}_1^T f_u^T)V_{xb}\\
-\dot{V}_{bb} &= -V_{xb}^T f_u Z^T(H_{uu}+\lambda\hat{g}_{uu})^{-1} Z f_u^T V_{xb}
\end{aligned} \tag{2.5.3}$$

$$-\dot{V}_{xx} = H_{xx} + f_x^T V_{xx} + V_{xx} f_x - (H_{ux} + f_u^T V_{xx})^T Z^T (H_{uu}+\lambda\hat{g}_{uu})^{-1} Z(H_{ux}+f_u^T V_{xx})$$

$$\hat{\beta}_1 = -(H_{uu}+\lambda\hat{g}_{uu})^{-1} Z(H_{ux}+f_u^T V_{xx}) \tag{2.5.4}$$

$$\hat{\beta}_2 = -(H_{uu} + \lambda \hat{g}_{uu})^{-1} Z f_u^T V_{xb} \tag{2.5.5}$$

$$u = \hat{u} + \hat{\beta}_1 \, \delta x + \hat{\beta}_2 \, \delta b \tag{2.5.6}$$

$$\delta b = -\varepsilon V_{bb}^{-1} V_b \Big|_{t_o} \tag{2.5.7}$$

λ is given by Equation (2.4.12).

2.5.2. The Computational Procedure

The procedure is the same as that of Section 2.3.2, except that H is minimized with respect to u for $u \in U$, which yields $\hat{g}$ and $\hat{u}$; λ is calculated using Equation (2.4.12).

2.5.3. Sufficient Conditions for an Improved Trajectory at Each Iteration

Sufficient conditions for $a(x_o, \bar{b}; t_o) < 0$ are the same as those of Section 2.4.4 for the free endpoint problem. In addition, for a reduction in terminal error, we require $V_{bb}(x_0, \bar{b}; t_0)$ to be invertible. Sufficient conditions are that:

(1) $(H_{uu} + \lambda \hat{g}_{uu})^{-1}$ be positive-definite
(2) $\hat{g}_u^T$ have full rank $\hat{p}$
(3) the linear system $\delta \dot{x} = (f_x + f_u \hat{\beta}_1) \delta x + f_u Z^T \delta u$ be completely controllable.

PROOF: The proof of these conditions is similar to that given in Section 2.3.4.

2.5.4. Characteristics of the Algorithm

In addition to the points listed in Section 2.4.3, the following is important: a necessary condition for invertibility of $V_{bb}(x_o, \bar{b}; t_o)$ is that the following linear system be completely controllable:

$$\delta \dot{x} = (f_x + f_u \hat{\beta}_1) \, \delta x + f_u Z^T \delta u \tag{2.5.8}$$

The controllability condition on Equation (2.5.8) may be very restrictive, depending on the type of constraints present and the nonoptimality of the nominal trajectory. Take for example a scalar control with constraint

$|u| \leqslant 1$. Obviously $Z = 1$ when the constraint is inoperative, and $Z = 0$ when it is operative. Now let us assume that for the particular nominal trajectory, and for the particular system,

$$\hat{u}(t) = \pm 1; \qquad t \in [t_o, t_f] \tag{2.5.9}$$

Clearly $Z = 0$ on the whole time interval $[t_o, t_f]$ and

$$V_{bb}(x_o, \bar{b}; t_o) = 0 \tag{2.5.10}$$

In this case one cannot use the algorithm because δb cannot be calculated, and, moreover, $u(t) = \hat{u}(t)$, which is independent of δb. One can overcome these difficulties by (1) using a first-order expression to compute δb, i.e., $\delta b = \varepsilon V_b$ ($\varepsilon > 0$), and (2) computing $u(t)$ using

$$\min_{u \in U} H(\bar{x} + \delta x, u, V_x + V_{xx}\delta x + V_{xb}\delta b; t)$$

After a few such iterations, V_{bb} may be invertible, at which stage one can return to the original algorithm.

2.6. SUMMARY

In this chapter we introduced the notion of global variations in control (strong variations in the x trajectory) into the differential dynamic programming method. We exploited this notion to develop new computational algorithms that are applicable to a larger class of nonlinear control problems than existing second-variation or successive sweep methods. In particular, problems with control inequality constraints, endpoint equality constraints, and implicitly given final time t_f were studied. We remark that mixed state control inequality constraints, Inequality (1.1.4), are amenable to the same treatment as that given in Section 2.4 for pure control inequality constraints. The extension is simple and is not given in this book.

In Chapter 3, strong variations in the x trajectory are used to study and solve a class of nonlinear bang-bang control problems.

References

1. D. H. Jacobson, *J. Opt. Theory Appl.* **2**, 411 (1968).
2. S. K. Mitter, *Automatica* **3**, 135 (1966).
3. S. R. McReynolds and A. E. Bryson, *Proc. 6th Joint Auto. Control Conf., Troy, New York, 1965*, p. 551.
4. D. Q. Mayne, *Intern. J. Control* **3**, 85 (1966).
5. S. R. McReynolds, *J. Math. Anal. Appl.* **19**, 565 (1967).
6. T. E. Bullock and G. F. Franklin, *IEEE Trans. Auto. Control* **AC-12**, 666 (1967).
7. H. Halkin, *in* "Computing Methods in Optimization Problems" (A. V. Balakrishnan and L. W. Neustadt, eds.), p. 211. Academic Press, New York, 1964.
8. R. Fletcher, *Comput. J.* **8**, 33 (1965).
9. R. Fletcher and M. J. D. Powell, *Comput. J.* **6**, 163 (1963).
10. D. H. Jacobson, *Program 008*, Control Section Program Library, Imperial College, Univ. of London, England 1966.
11. R.E. Kalman, *Bol. Soc. Mat. Mex.* **5**, 102 (1960).
12. S. B. Gershwin and D. H. Jacobson, Harvard Univ. Tech. Rep. **TR 566** (August 1968).
13. S. E. Dreyfus, Rand Rept. **P-3487** (1966).
14. J. R. Gantmacher, "Matrix Theory," Vol. 1, p. 247. Chelsea, New York, 1959.
15. S. E. Dreyfus, Rand Rept. **P-2605** (1962).
16. A. R. M. Noton, P. Dyer, and C. A. Markland, *IEEE Trans. Auto. Control* **AC-12**, 59 (1967).
17. G. Hadley, "Non-Linear and Dynamic Programming," Addison-Wesley, Reading, Massachusetts, 1964.
18. G. P. McCormick, *SIAM J. Appl.Math.* **15**, 641 (1967).

Chapter 3

NEW ALGORITHMS FOR THE SOLUTION OF BANG-BANG CONTROL PROBLEMS

3.1. INTRODUCTION

In the previous chapter we showed some advantages that can be gained by allowing global variations in control. In particular control constrained problems of a certain class were solved.

In this chapter we further exploit the notion of global variations in control (strong variations in x) in order to solve the rather special class of bang-bang control problems. Besides producing new algorithms, the approach yields new insight into the structure of the optimal control and optimal cost functions.

3.2. A CLASS OF BANG-BANG CONTROL PROBLEMS

3.2.1. Formulation and Preliminary Investigation

We shall consider the dynamic system described by the following set of ordinary differential equations:

$$\dot{x} = f(x, u; t) = f_1(x; t) + f_2(x; t)u; \qquad x(t_o) = x_o \tag{3.2.1}$$

where f_1 is an n-dimensional, nonlinear, vector function of x and t, f_2 is an $n \times m$ matrix function x and t; and u is an m-dimensional control vector.

We assume that the controls u_j; $j = 1, \ldots, m$ are constrained in the following way:

$$u_j^b \leqslant u_j \leqslant u_j^a; \qquad j = 1, \ldots, m \tag{3.2.2}$$

where u_j^b and u_j^a are constants.

The problem is to choose $u(t)$; $t \in [t_o, t_f]$ to satisfy Constraints (3.2.2) and to minimize

$$V(x_o; t_o) = \int_{t_o}^{t_f} L(x; t)\, dt + F(x(t_f); t_f) \tag{3.2.3}$$

Forming the Hamiltonian, we have

$$H(x, u, V_x; t) = L(x; t) + \langle V_x, f_1(x; t) + f_2(x; t)u \rangle \qquad (3.2.4)$$

H is minimized with respect to u if

$$\left.\begin{aligned} u_j &= u_j^b; \qquad (f_2^T V_x)_j > 0 \\ u_j &= u_j^a; \qquad (f_2^T V_x)_j < 0 \end{aligned}\right\} \quad j = 1, \ldots, m \qquad (3.2.5)$$

It is assumed that $(f_2^T V_x)_j$; $j = 1, \ldots, m$ is not zero on a finite time interval† because otherwise one could not determine the minimizing u_j from Equation (3.2.4), since H would be insensitive to u_j.

The control law, Equations (3.2.5) is referred to as bang-bang because the u_j switch between their upper and lower limits depending on the sign of $(f_2^T V_x)_j$. It is assumed that the control functions have a finite number of switch times t_{s_i}; $i = 1, \ldots, n_s$.

The following question arises when attempting to study the above problem using dynamic programming: do the partial derivatives of V^o with respect to x and t exist everywhere in the state space, and if not, how does one use the Bellman PDE, which is derived from the assumption that the optimal cost V^o has continuous first-partial and second-partial derivatives with respect to x and t?

Consider Equation (3.2.3) and assume that $u_j = u_j^a$; $j = 1, \ldots, m$, for the whole time interval $[t_o, t_f]$; $V(x; t)$ will have continuous partial derivatives throughout the state space if L and f are sufficiently well-behaved functions‡.

Now consider the case where the controls u_j; $j = 1, \ldots, m$, are given by Equations (3.2.5). At times when $(f_2^T V_x)_j$, say, changes sign, the controls u_j will jump discontinuously from u_j^b to u_j^a or vice versa. It is not clear whether $V(x; t)$ will have continuous partial derivatives at such points of discontinuous control.

In view of the above, we can conclude that if the u_j are constant, then Bellman's PDE is valid. If there are switch times (t_{s_i}) present where the u_j change discontinuously, one must investigate further to discover exactly what happens to $V(x; t)$ and its partial derivatives at these times.

---

† If this assumption is violated, the problem is called singular. Singular problems have been studied [1-4], but general computational procedures are lacking.

‡ It is easy to show that, for constant u_j; $j = 1, \ldots, m$, the partial derivatives of V satisfy linear differential equations that have bounded solutions in the time interval $[t_o, t_f]$.

3.3. A NEW SECOND-ORDER ALGORITHM FOR FREE ENDPOINT BANG-BANG CONTROL PROBLEMS †

3.3.1. Piecewise Differential Dynamic Programming

Assume that we have a nominal control $\bar{u}(t)$; $t \in [t_o, t_f]$, that satisfies the Constraints (3.2.2). A nominal trajectory $\bar{x}(t)$; $t \in [t_o, t_f]$ is produced using this control.

From Equation (1.4.13) the PDE satisfied by the local, second-order expansion for the cost function is: ‡

$$-\frac{\partial \bar{V}}{\partial t} - \frac{\partial a}{\partial t} - \left\langle \frac{\partial V_x}{\partial t}, \delta x \right\rangle - \tfrac{1}{2}\left\langle \delta x, \frac{\partial V_{xx}}{\partial t}\delta x \right\rangle$$

$$= \min_{\delta u} \; [L(\bar{x},+\delta x; t) + \langle V_x + V_{xx}\delta x, f(\bar{x}+\delta x, \bar{u}+\delta u; t)\rangle] \qquad (3.3.1)$$

where

$$f(\bar{x}+\delta x, \bar{u}+\delta u; t) = f_1(\bar{x}+\delta x; t) + f_2(\bar{x}+\delta x; t)(\bar{u}+\delta u) \qquad (3.3.2)$$

Setting $\delta x = 0$ in Equation (3.3.1), we obtain

$$-(\partial \bar{V}/\partial t - \partial a/\partial t) = \min_{\delta u} [L(\bar{x}; t) + \langle V_x, f(\bar{x}, \bar{u}+\delta u; t)\rangle] \qquad (3.3.3)$$

Now let us carry out the minimization in Equation (3.3.3) using the rules of Equations (3.2.5). Let us assume that the minimizing u^* is

$$u^* = \bar{u} + \delta u^* = u^+ \qquad (3.3.4)$$

where the components of u^+ are given by Equations (3.2.5). Now let us reintroduce δx; i.e., we again consider state $x = \bar{x}+\delta x$. § We shall assume that the minimizing control for $\bar{x}+\delta x$ still remains as u^+, except in the neighborhood of switch points of u^+. So, from Equation (3.3.1):

$$-\frac{\partial \bar{V}}{\partial t} - \frac{\partial a}{\partial t} - \left\langle \frac{\partial V_x}{\partial t}, \delta x \right\rangle - \tfrac{1}{2}\left\langle \delta x, \frac{\partial V_{xx}}{\partial t}\delta x \right\rangle = L(\bar{x}+\delta x; t)$$

$$+ \langle V_x + V_{xx}\delta x, f(\bar{x}+\delta x, u^+; t)\rangle \qquad (3.3.5)$$

---

† Jacobson [5].

‡ Again, $V_{xxx}\delta x \delta x$ is neglected *a priori* to avoid having to carry this throughout the derivation. An error analysis, somewhat different from that of Section 2.2.1 is given in Appendix B.

§ c.f. Section 2.2.2.

The right-hand side of Equation (3.3.5) expanded to second-order in δx is

$$H + \langle H_x, \delta x\rangle + \langle V_{xx} f, \delta x\rangle + \tfrac{1}{2}\langle \delta x, (H_{xx} + f_x^T V_{xx} + V_{xx} f_x)\delta x\rangle \tag{3.3.6}$$

All quantities are evaluated at $\bar{x}$ and u^+.

In the same way as described in Section 2.2.1, we obtain

$$-\dot{a} = H - H(\bar{x}, \bar{u}, V_x; t) \tag{3.3.7}$$

$$-\dot{V}_x = H_x + V_{xx}(f - f(\bar{x}, \bar{u}; t)) \tag{3.3.8}$$

$$-\dot{V}_{xx} = H_{xx} + f_x^T V_{xx} + V_{xx} f_x \tag{3.3.9}$$

All quantities are evaluated at $\bar{x}$ and u^+ unless stated otherwise.

At $t = t_f$, we have from Equation (3.2.3), that

$$a(t_f) = 0 \tag{3.3.10}$$

$$V_x(t_f) = F_x(\bar{x}(t_f); t_f) \tag{3.3.11}$$

$$V_{xx}(t_f) = F_{xx}(\bar{x}(t_f); t_f) \tag{3.3.12}$$

Equations (3.3.7) to (3.3.9) can be integrated backward from t_f using the boundary conditions given by Equations (3.3.10) to (3.3.12), and $u^* = u^+$.

There will probably come a time t_s where a component of u^* changes discontinuously (i.e., switches) so that u^* changes from u^+ to u^-, say.†
Conditions at t_s must now be studied.

3.3.2. Conditions at Switch Points of the Control Function

Let the cost for $t \geqslant t_s$ be denoted by

$$V^+(\bar{x}+\delta x; t) = \bar{V}^+ + a^+ + \langle V_x^+, \delta x\rangle + \tfrac{1}{2}\langle \delta x, V_{xx}^+ \delta x\rangle \tag{3.3.13}$$

For $t \in [t_0, t_s]$, the cost is given by

$$V^-(\bar{x}(t) + \delta x(t); t) = \int_t^{t_s} L(\bar{x}(\tau) + \delta x(\tau); \tau)\, d\tau + V^+(\bar{x}(t_s) + \delta x(t_s); t_s) \tag{3.3.14}$$

† In general, if more than one component (say r, where $r \leqslant m$) of u^* changes at the same time t_s, it may be necessary to consider the separate changes in switch times δt_s^1, δt_s^2, ..., δt_s^r, for each control, in the analysis to follow. In this chapter we treat the case where only one component of the control switches at t_s.

where superscript $-$ on V denotes the cost for $t \leqslant t_s$ in the region where $u^* = u^-$.

One can consider V^- to be a function of the switch time t_s; so let us consider t_s to be a parameter of V^- and write it explicitly as such:

$$V^-(\bar{x}(t) + \delta x(t), t_s; t) = \int_t^{t_s} L(\bar{x}(\tau) + \delta x(\tau); \tau)\, d\tau + V^+(\bar{x}(t_s) + \delta x(t_s); t_s) \tag{3.3.15}$$

Now we allow variations δt_s in the switch time:

$$V^-(\bar{x}(t) + \delta x(t), t_s + \delta t_s; t) = \int_t^{t_s+\delta t_s} L(\bar{x}(\tau) + \delta x(\tau); \tau)\, d\tau + V^+(\bar{x}(t_s+\delta t_s) + \delta x(t_s+\delta t_s); t_s+\delta t_s) \tag{3.3.16}$$

Writing $\bar{x}(\tau) + \delta x(\tau)$ and $\bar{x}(t_s + \delta t_s) + \delta x(t_s+\delta t_s)$ in terms of $\bar{x}(t) + \delta x(t)$, Equation (3.3.16) becomes

$$V^-(\bar{x}(t) + \delta x(t), t_s + \delta t_s; t) = \int_t^{t_s+\delta t_s} L(\bar{x}(t) + \delta x(t) + \Delta x(\tau); \tau)\, d\tau + V^+(\bar{x}(t) + \delta x(t) + \Delta x(t_s+\delta t_s); t_s+\delta t_s) \tag{3.3.17}$$

where

$$\Delta x(\tau) = \int_t^{\tau} f(\bar{x}(\tau_1) + \delta x(\tau_1), u^-; \tau_1)\, d\tau_1 \tag{3.3.18}$$

Conditions are to be studied at $t = t_s$; so let us observe Equation (3.3.17) at $t = t_s$:†

$$V^-(\bar{x}(t_s) + \delta x(t_s), t_s+\delta t_s; t_s) = \int_{t_s}^{t_s+\delta t_s} L(\bar{x}(t_s) + \delta x(t_s) + \Delta x(\tau); \tau)\, d\tau + V^+(\bar{x}(t_s) + \delta x(t_s) + \Delta x(t_s+\delta t_s); t_s+\delta t_s) \tag{3.3.19}$$

† $V^-(\bar{x}(t_s)+\delta x(t_s), t_s+\delta t_s; t_s)$ is the cost, at time $t = t_s$, as a function of the parameters $\bar{x}(t_s)+\delta x(t_s)$ and $t_s+\delta t_s$. The semicolon separates these parameters from the actual time $t = t_s$ at which V^- is observed. Thus, $V_{t_s}^-(\bar{x}(t), t_s; t)$ is the partial derivative of V^- with respect to the parameter t_s, evaluated at $\bar{x}(t), t_s; t$. So $V_{t_s}^-(\bar{x}(t_s), t_s; t_s)$ is this partial derivative observed at time t_s. $V_t^-(\bar{x}(t), t_s; t)$ is the partial derivative of V^- with respect to the time t, evaluated at $\bar{x}(t), t_s; t$. So $V_t^-(\bar{x}(t_s), t_s; t_s)$ is this partial derivative observed at time $t = t_s$. For convenience, and where the meaning is clear, $\bar{x}(t)$ will be written as $\bar{x}$.

Equation (3.3.19) is precisely the same form as Equation (2.3.52), for the implicit final-time problem, with ψ absent and V^+ replacing F.

Expanding the left-hand side of Equation (3.3.19) to second-order about $\bar{x}, t_s$:

$$V^-(\bar{x}, t_s; t_s) + \langle V_x^-, \delta x\rangle + V_{t_s}^- \delta t_s + \langle V_{xt_s}^-, \delta x\rangle \delta t_s + \tfrac{1}{2}\langle \delta x, V_{xx}^- \delta x\rangle + \tfrac{1}{2} V_{t_s t_s}^- \delta t_s^2 \tag{3.3.20}$$

The expansion of the right-hand side of Equation (3.3.19) is the same as that given in Section 2.3.5. Equating coefficients of like powers of δx and δt_s, one obtains

$$\begin{aligned} V^-(\bar{x}, t_s; t_s) &= V^+(\bar{x}; t_s) \\ V_x^- &= V_x^+ \\ V_{xx}^- &= V_{xx}^+ \\ V_{t_s}^- &= H^- + V_t^+ \\ V_{xt_s}^- &= V_{xt}^+ + H_x^- + V_{xx}^+ f^- \\ V_{t_s t_s}^- &= H_t^- + V_{tt}^+ + \langle H_x^-, f^-\rangle + 2\langle V_{xt}^+, f^-\rangle + \langle f^-, V_{xx}^+ f^-\rangle \end{aligned} \tag{3.3.21}$$

where

$$H^{\mp} = H(\bar{x}, u^{\mp}, V_x^+; t_s) \tag{3.3.22}$$

From Equations (3.3.5) and (3.3.6),

$$\begin{aligned} V_t^+ &= \bar{V}_t^+ + a_t^+ = -H^+ \\ V_{xt}^+ &= -H_x^+ - V_{xx}^+ f^+ \\ V_{tt}^+ &= -H_t^+ + \langle H_x^+, f^+\rangle + \langle f^+, V_{xx}^+ f^+\rangle \end{aligned} \tag{3.3.23}$$

Substituting Equations (3.3.23) into Equations (3.3.21), we obtain

$$V^-(\bar{x}, t_s; t_s) = V^+(\bar{x}; t_s)$$

$$V_x^- = V_x^+$$

$$V_{xx}^{-} = V_{xx}^{+}$$

$$V_{t_s}^{-} = H^{-} - H^{+} \tag{3.3.24}^{\dagger}$$

$$V_{xt_s}^{-} = H_x^{-} - H_x^{+} + V_{xx}^{+}(f^{-} - f^{+})$$

$$V_{t_s t_s}^{-} = H_t^{-} - H_t^{+} - \langle H_x^{-}, f^{-} - f^{+} \rangle + \langle f^{-}, H_x^{-} - H_x^{+} \rangle$$
$$+ \langle f^{-} - f^{+}, V_{xx}^{+}(f^{-} - f^{+}) \rangle$$

The quantities in Equation (3.3.24) are sensitivities of the cost V^{-} with respect to the parameters x and t_s at time $t = t_s$. From Equation (3.3.24),

$$V_{t_s}^{-} = H^{-} - H^{+} = \langle V_x^{+}, f_2[u^{-} - u^{+}] \rangle \tag{3.3.25}$$

From Equation (3.3.25),

$$V_{t_s}^{-} = 0 \tag{3.3.26}$$

since either

$$u_j^{-} = u_j^{+} \qquad \text{or} \quad (f_2^T V_x)_j = 0 \tag{3.3.27}$$

Equation (3.3.26) is a necessary condition for V^{-} to be minimized with respect to the switch time t_s. A further necessary condition is

$$V_{t_s t_s}^{-} \geqslant 0 \tag{3.3.28}$$

From Equation (3.3.20),

$$(\partial V^{-}/\partial t_s)(\bar{x} + \delta x,\, t_s + \delta t_s\,;\, t_s) = V_{t_s}^{-} + \langle V_{xt_s}^{-}, \delta x \rangle + V_{t_s t_s}^{-} \delta t_s \tag{3.3.29}$$

In order to maintain the necessary condition of optimality [Equation (3.3.26)] for the case where variations δx are present, it is required that

$$V_{t_s}^{-}(\bar{x} + \delta x,\, t_s + \delta t_s\,;\, t_s) = 0 \tag{3.3.30}$$

Using Equations (3.3.26), (3.3.29), and (3.3.30),

$$\delta t_s = -V_{t_s t_s}^{-1} \langle V_{xt_s}^{-}, \delta x \rangle \tag{3.3.31}$$

† Obtained in an equivalent form, independently, by Dyer and McReynolds [6].

Equation (3.3.31) is an optimal, local linear feedback controller relating the required change in switch time δt_s to the δx appearing at $t = t_s$.

Substituting Equation (3.3.31) into Equation (3.3.20) to eliminate δt_s, we obtain

$$V^-(\bar{x}+\delta x, t_s-(V^-_{t_s t_s})^{-1}\langle V^-_{xt_s}, \delta x\rangle; t_s) = V^-(\bar{x}, t_s; t_s) + \langle V_x^-, \delta x\rangle$$

$$+\tfrac{1}{2}\langle \delta x, (V^-_{xx} - (V^-_{xt_s}\cdot V^-_{t_s x})/V^-_{t_s t_s})\delta x\rangle \qquad (3.3.32)$$

Renaming the left-hand side of Equation (3.3.32) as $\hat{V}(\bar{x}+\delta x; t_s)$ and using Equations (3.3.24), the following expressions result:

$$\hat{V}(\bar{x}; t_s) = V^+(\bar{x}; t_s)$$

$$\hat{V}_x = V_x^+ \qquad (3.3.33)$$

$$\hat{V}_{xx} = V^+_{xx} - (V^-_{xt_s}\cdot V^-_{t_s x})/V^-_{t_s t_s}$$

With no loss of correctness, the superscripts may now be dropped to yield:

$$\hat{V}(\bar{x}; t_s) = V(\bar{x}; t_s)$$

$$\hat{V}_x = V_x \qquad (3.3.34)$$

$$\hat{V}_{xx} = V_{xx} - (V_{xt_s}\cdot V_{t_s x})/V_{t_s t_s}$$

The following points are noteworthy: (1) At $t = t_s$, $a(\bar{x}; t_s)$ is continuous, since $\hat{V} = V$ and $\bar{V}(\bar{x}; t)$ is continuous. (2) At $t = t_s$, V_x is continuous. (3) At $t = t_s$, V_{xx} experiences a jump of magnitude

$$\Delta V_{xx} = -(V_{xt_s}\cdot V_{t_s x})/V_{t_s t_s} \qquad (3.3.35)$$

At a switch point the jump in V_{xx} is readily computed using Equations (3.3.24) and (3.3.35). Equations (3.3.34) then become new boundary conditions for the differential equations (3.3.7) to (3.3.9), which can continue to be integrated backward, noting now that $u^* = u^-$.

Note that if $u^+ = u^-$, and if $\dot{u}^+ \neq \dot{u}^-$, then, from Equations (3.3.24), V is unaffected, to second-order by a change in t_s. This is the case, described in Section 2.4.1, where the optimal control on entering or leaving a constraint boundary is assumed to be continuous.

The above analysis is clearly applicable to a control function u^* having a finite number of switch times. The above theory was developed for non-optimal $\bar{x}$, $\bar{u}$. On an optimal trajectory, a special case, all the results hold.

3.3.3. An Example

Consider the following control problem:

$$\begin{aligned} \dot{x}_1 &= x_2; & x_1(t_o) &= x_{1o} \\ \dot{x}_2 &= u\ ; & x_2(t_o) &= x_{2o} \end{aligned} \tag{3.3.36}$$

$$|u| \leqslant 1$$

Minimize

$$V = \int_{t_o}^{\infty} x_1^2\, dt \tag{3.3.37}$$

Fuller [11] has analytic expressions for the optimal cost $V^o(x_1, x_2)$. (V^o is independent of t because $t_f = \infty$.) The question now is whether the predictions, that $V_x{}^o$ is continuous at a switch point and V_{xx}^o experiences a jump, hold for this problem.

Differentiation of the analytic expressions for $V^o(x_1, x_2)$ does indeed confirm these predictions, and moreover, the jump in V_{xx}^o agrees with that predicted by Equation (3.3.35). For this problem,

$$H(x, u, V_x; t) = x_1^2 + V_{x_1} x_2 + V_{x_2} u \tag{3.3.38}$$

where the minimizing control is

$$u^* = -\text{sign}\, V_{x_2} \tag{3.3.39}$$

(The "optimal superscript" is dropped, for convenience.)

Fuller has obtained the following expressions for the optimal cost: $V^N(x_1, x_2)$—the cost surface when $u = -1$—is given by

$$V^N(x_1, x_2) = x_1^2 x_2 - \tfrac{2}{3} x_1 x_2^3 + \tfrac{2}{15} x_2^5 + 0.764(x_1 + \tfrac{1}{2} x_2^2)^{5/2} \tag{3.3.40}$$

$V^P(x_1, x_2)$— the cost surface when $u = +1$—is given by

$$V^P(x_1, x_2) = -x_1^2 x_2 + \tfrac{2}{3} x_1 x_2^3 - \tfrac{2}{15} x_2^5 + 0.764(-x_1 + \tfrac{1}{2} x_2^2)^{5/2} \tag{3.3.41}$$

On the switching curve,

$$x_1 = -0.4446\, x_2{}^2; \qquad x_2 > 0$$
$$x_1 = 0.4446\, x_2{}^2; \qquad x_2 < 0 \tag{3.3.42}$$

Consider only the case where $x_2 < 0$. Substituting Equation (3.3.42) into Equations (3.3.40) and (3.3.41),

$$V^N(0.4446x_2{}^2, x_2) = V^P(0.4446x_2{}^2, x_2) = -0.035\, x_2{}^5 \tag{3.3.43}$$

i.e., V is continuous at a switch point of the control.

From Equations (3.3.40) and (3.3.41),

$$V^N_{x_1}(x_1, x_2) = 2x_1x_2 + \tfrac{2}{3}x_2{}^3 + \tfrac{5}{2}\cdot 0.764(x_1 + \tfrac{1}{2}x_2{}^2)^{3/2} \tag{3.3.44}$$

$$V^P_{x_1}(x_1, x_2) = -2x_1x_2 + \tfrac{2}{3}x_2{}^3 - \tfrac{5}{2}\cdot 0.764(-x_1 + \tfrac{1}{2}x_2{}^2)^{3/2} \tag{3.3.45}$$

Using Equations (3.3.42), (3.3.44), and (3.3.45),

$$V^N_{x_1} = V^P_{x_1} = -0.2\, x_2{}^3 \qquad \text{at a switch point} \tag{3.3.46}$$

i.e., V_{x_1} is continuous at a switch point. It can be easily shown that V_{x_2} is also continuous.

$$V^N_{x_1x_1}(x_1, x_2) = 2x_2 + \tfrac{15}{4}\cdot 0.764(x_1 + \tfrac{1}{2}x_2{}^2)^{1/2}$$
$$= -0.79\, x_2 \qquad \text{at a switch point} \tag{3.3.47}$$

$$V^P_{x_1x_1}(x_1, x_2) = -2x_2 + \tfrac{15}{4}\cdot 0.764(-x_1 + \tfrac{1}{2}x_2{}^2)^{1/2}$$
$$= -2.67\, x_2 \qquad \text{at a switch point} \tag{3.3.48}$$

So crossing the switch point along an optimal path from N to P, $V_{x_1x_1}$ experiences a jump of

$$\Delta V_{x_1x_1} = -0.79\, x_2 + 2.67\, x_2 = 1.88\, x_2 \tag{3.3.49}$$

Consider the expression for V_{xt_s} and $V_{t_s t_s}$ given by Equation (3.3.24):

$$V_{xt_s} = V^P_{xx}(f^N - f^P) = -2\,V^P_{xx} \cdot \begin{bmatrix} 0 \\ 1 \end{bmatrix} \tag{3.3.50}$$

So

$$V_{xt_s} \cdot V_{t_s x} = 4 \begin{bmatrix} V^2_{x_1 x_2} & V_{x_1 x_2} \cdot V_{x_2 x_2} \\ V_{x_2 x_2} \cdot V_{x_1 x_2} & V^2_{x_2 x_2} \end{bmatrix} \tag{3.3.51}$$

$$\begin{aligned} V_{t_s t_s} &= -H_x{}^P(f^N - f^P) + \langle f^N - f^P, V^P_{xx}(f^N - f^P)\rangle \\ &= 2\,V^P_{x_1} + 4\,V^P_{x_2 x_2} \end{aligned} \tag{3.3.52}$$

Now

$$\begin{aligned} V^P_{x_1 x_2} &= -2x_1 + 2x_2{}^2 - \tfrac{15}{4} \cdot 0.764(-x_1 + \tfrac{1}{2}x_2{}^2)^{1/2} \\ &= 1.785\,x_2{}^2 \qquad \text{at a switch point} \end{aligned} \tag{3.3.53}$$

and

$$\begin{aligned} V^P_{x_2 x_2} &= 4x_1 x_2 - \tfrac{8}{3}x_2{}^3 + 0.764(-x_1 + \tfrac{1}{2}x_2{}^2)^{1/2} x_2{}^2 \\ &\quad + \tfrac{15}{4} \cdot 0.764(-x_1 + \tfrac{1}{2}x_2{}^2)^{3/2} = -1.59\,x_2{}^3 \qquad \text{at a switch point} \end{aligned} \tag{3.3.54}$$

Substituting Equations (3.3.53) and (3.3.54) into Equations (3.3.51) and (3.3.52), and using Equation (3.3.35),

$$\Delta V_{x_1 x_1} = -(1.785\,x_2{}^2)^2/(-1.59\,x_2{}^3 - 0.1\,x_2{}^3) = [(1.785)^2/1.69]\,x_2 = 1.88\,x_2 \tag{3.3.55}$$

The predicted jump given by Equation (3.3.55) is the same as the actual jump given by Equation (3.3.49). Similar agreement can be obtained for the other elements of V_{xx}, and for the parts of the state space where $x_2 > 0$.

3.3.4. Implementation of the New Control

The parameters $a(t)$, $V_x(t)$, and $V_{xx}(t)$ of the quadratic expansion for $V(\bar{x} + \delta x;\, t)$ are easily computed backward along a nominal trajectory using Equations (3.3.7) to (3.3.12), and Equations (3.3.24) and (3.3.34).† What is

† It is assumed that t_f is given explicitly.

now required is a method of using this data to improve the current nominal trajectory. A procedure that comes immediately to mind is the following: apply a new control function computed from the expression

$$u_j(t) = -\operatorname{sign}[f_2^T(\bar{x}+\delta x; t)(V_x + V_{xx}\delta x)]_j; \qquad j = 1, \ldots, m \tag{3.3.56}$$

However, this procedure is unsound because $V_{xx}(t)$ experiences a jump at each switch time of $u^*(t)$; thus $V_x + V_{xx}\delta x$ experiences a jump at switch times of $u^*(t)$ and this may cause $u(t)$ to switch when it should not.‡

The above difficulty is overcome by using the local, linear controller, Equation (3.3.31) to compute changes in the switch times. The controller is given below in general form:

$$\delta t_{s_i} = -V_{t_s t_s}^{-1}(\bar{x}, t_{s_i}; t_{s_i})\langle V_{xt_s}(\bar{x}, t_{s_i}; t_{s_i}), \delta x(t_{s_i})\rangle \tag{3.3.57}$$

where t_{s_i} is the time of the ith switch of the control $u^*(t)$.

Equation (3.3.57) is used in the following way: run forward in time using the new controls,

$$u_j(t) = u_j^*(t); \qquad j = 1, \ldots, m \tag{3.3.58}$$

When a switch time t_s, say, of a control $u_k(t) = u_k^*(t)$; $k \in \{1, \ldots, m\}$ is reached, measure $\delta x(t_s)$ and calculate δt_s using Equation (3.3.57). If $\delta t_s > 0$, hold $u_k(t) = u_k^*(t_s^-)$§ for the time interval δt_s; after this, once again set $u_k(t) = u_k^*(t)$ and continue. If, however, $\delta t_s < 0$, then backspace the integration routine by the amount δt_s and starting at this time $t_s - \delta t_s$, set $u_k(\tau) = u_k^*(t_s^+)$ for $t_s - \delta t_s \leqslant \tau \leqslant t_s^+$ and integrate forward again. After time $t = t_s^+$, once again set $u_k(t) = u_k^*(t)$ and continue.

The above procedure implements the local feedback controller, Equation (3.3.57), directly; there is thus no chance of discontinuities appearing in $u(t)$ where there should be none, as happens if Equation (3.3.56) is used.

---

‡ On the forward run, the position of the switch times of $u(t)$ should not coincide with those of $u^*(t)$, unless $u^*(t)$ happens to be optimal. Friedland and Sarachik [14] used Equation (3.3.56) for local feedback control. The fact that it is an unsound scheme is illustrated by their numerical results which show numerous switching in the control functions for examples where there should be only one switch.

§ In this section the plus and minus superscripts denote the time instants immediately before and after t_s, respectively.

Applying the new control on the whole time interval $[t_o, t_f]$ may produce a δx that is too large.† The "step size adjustment method" of Section 2.2.2 must then be used to limit the size of δx to suitable values.

3.3.5. The Computational Procedure

The computational procedure is much the same as that described in Section 2.2.3; however, on the forward run, the new control is computed using the procedure of Section 3.3.4.

3.3.6. Characteristics of the Algorithm

1. It is believed that this is the first time that dynamic programming has been used to derive a second-order algorithm for solving bang-bang control problems.‡ The approach explicitly uses the bang-bang structure to find a local, linear feedback controller, between switch time changes and small differences in the state from the nominal. This controller is then used to generate a new, improved trajectory.

2. If a nominal control is given whose switch times are sufficiently close to the optimal ones, then convergence is quadratic in the sense that Equation (3.3.57) is valid.

3.3.7. Sufficient Conditions for a Reduction in Cost at Each Iteration

It is assumed in this section that $\bar{u}(t)$; $t \in [t_o, t_f]$ satisfies the control constraints.

For $t \in [t_o, t_f]$;

1. $[f_2^T(\bar{x}; t) \cdot V_x(\bar{x}; t)]_j \neq 0$ on a finite time interval; $j = 1, \ldots, m$.
2. Except at switch points,
$$H(\bar{x}, u^*, V_x; t) < H(\bar{x}, \bar{u}, V_x; t); \qquad u^* \neq \bar{u}$$
3. $V_{t_s t_s}(\bar{x}, t_{s_i}; t_{s_i}) > 0; \qquad i = 1, \ldots, n_s.$
4. The solutions of the differential equations (3.3.7) to (3.3.9) be bounded.

---

† The theory of Sections 3.3.1 and 3.3.2 is based on the assumption that δx is sufficiently small to justify the use of second-order expansions for V, L, and f with respect to x.

‡ Recently, and independently, Dyer and McReynolds [6] have obtained similar results to those described in this chapter. However, the algorithms presented in this chapter are simpler to implement, in that the nominal, assumed control function is not required to contain an *a priori* specified number of switchings (in fact, the nominal control need not even be of bang-bang form).

PROOF: In order that u^* be determined by Equations (3.2.5), it is necessary that (1) be satisfied; otherwise H would be insensitive to u_j on a finite time interval.

For the jumps in V_{xx} at $t_{s_i}(i=1, \dots, n_s)$ to be bounded, we must have

$$V_{t_s t_s}(\bar{x}, t_{s_i}; t_{s_i}) \neq 0 \tag{3.3.59}$$

Further, since V is to be minimized with respect to the switch times t_{s_i}, a necessary condition of optimality is

$$V_{t_s t_s}(\bar{x}, t_{s_i}; t_{s_i}) \geqslant 0 \tag{3.3.60}$$

Equations (3.3.59) and (3.3.60) yield condition (3).

The predicted reduction in cost is

$$a(x; t_1) = \int_{t_f}^{t_1} [H(\bar{x}, u^*, V_x; t) - H(\bar{x}, \bar{u}, V_x; t)]\, dt \tag{3.3.61}$$

A sufficient condition for the negativity of $a(x; t_1)$ is clearly that, except at switch points,

$$H(\bar{x}, u^*, V_x; t) < H(\bar{x}, \bar{u}, V_x; t); \qquad u^* \neq \bar{u} \tag{3.3.62}$$

In order that these quantities be bounded, the solutions of Equations (3.3.7) to (3.3.9) must be bounded.

3.3.8. A New First-Order Algorithm for Free Endpoint Bang-Bang Control Problems

A first-order algorithm is easily derived if the expansion, given by Equation (1.4.10), is truncated after the first-order terms:

$$V(\bar{x}+\delta x; t) = \bar{V} + a + \langle V_x, \delta x \rangle \tag{3.3.63}$$

The following equations result:

$$\begin{aligned} -\dot{a} &= H - H(\bar{x}, \bar{u}, V_x; t); && a(t_f) = 0 \\ -\dot{V}_x &= H_x; && V_x(t_f) = F_x(\bar{x}(t_f); t_f) \end{aligned} \tag{3.3.64}$$

where the above quantities are evaluated at $\bar{x}$, $\bar{u}^*$ unless stated otherwise; u^* is given by Equations (3.2.5), as before.

The new control is given by

$$u(t) = u^*(t); \qquad t \in [t_o, t_f] \tag{3.3.65}$$

The "step size adjustment method" of Section 2.2.2. is used to ensure that δx remains small enough.

The algorithm is the same as the first-order method described in Section 2.2.7; it is, therefore, not discussed further.

3.3.9. Computed Examples

The computed examples in this section serve to illustrate the usefulness of the second-order algorithms.

EXAMPLE I: The example is that of Section 3.3.3. However, the upper limit of integration in Equation (3.3.37) is taken as three seconds.

$$V = \int_0^3 x_1^2 \, dt \tag{3.3.66}$$

Initial conditions for Equations (3.3.36) are

$$x_1(o) = 1, \qquad x_2(o) = 0 \tag{3.3.67}$$

The new second-order algorithm of Section 3.3 was programmed; a fourth-order Runge-Kutta integration routine was used, and the interval [0, 3] was divided into 500 steps. A nominal control

$$\bar{u}(t) = +1; \qquad t \in [t_o, t_f] \tag{3.3.68}$$

was used; this produced a nominal cost of 12.1.

Figure 3.1 illustrates the cost as a function of iteration number. The reduction from 12.1 to the optimal value of 0.383 was accomplished in five iterations. Figure 3.2, shows phase plane portraits for some iterations, illustrating the movement of switch points, from iteration to iteration.

EXAMPLE II: This example is one tried by Plant and Athans [7] using a boundary-value iteration method. Plant and Athans considered the problem of hitting the unit sphere centered on the origin of the state space of a linear system in minimum time.

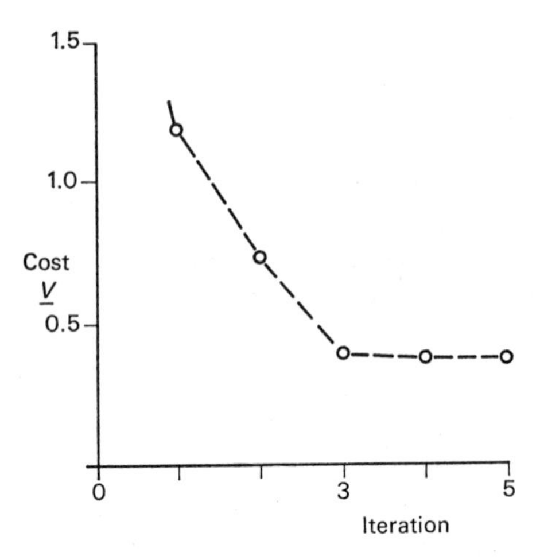

Figure 3.1. Second-order system: cost versus iteration number.

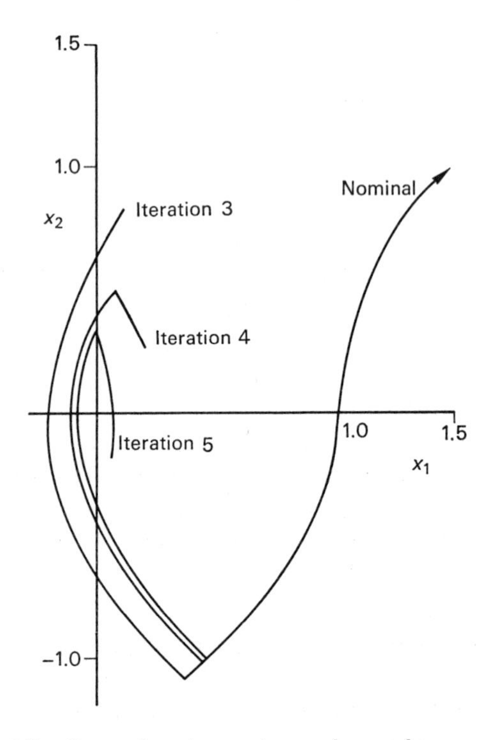

Figure 3.2. Second-order system: phase-plane portraits.

Here it is convenient to consider the following problem formulation:

$$
\begin{aligned}
\dot{x}_1 &= -0.5x_1 + 5x_2; & x_1(o) &= 10 \\
\dot{x}_2 &= -5x_1 - 0.5x_2 + u; & x_2(o) &= 10 \\
\dot{x}_3 &= -0.6x_3 + 10x_4; & x_3(o) &= 10 \\
\dot{x}_4 &= -10x_3 - 0.6x_4 + u; & x_4(o) &= 10
\end{aligned}
\tag{3.3.69}
$$

$$|u| \leqslant 1$$

Minimize

$$V = \langle x(t_f), x(t_f) \rangle; \qquad t_f = 4.2 \text{ sec} \tag{3.3.70}$$

A Runge-Kutta fourth-order integration routine was used with 300 integration steps. A nominal control

$$\bar{u}(t) = +1; \qquad t \in [0, 4.2] \tag{3.3.71}$$

was used and this produced a cost of 4.12. The second-order algorithm reduced this cost to the minimum value of 0.996 in two iterations. Figure 3.3 shows the cost as a function of iteration number and Figure 3.4 shows the nominal and optimal control functions; note the difference in structure between these two control functions.

3.4. A NEW SECOND-ORDER ALGORITHM FOR FIXED ENDPOINT BANG-BANG CONTROL PROBLEMS

3.4.1. The Derivation

In this section, constraints of the following form are treated:

$$\psi(x(t_f); t_f) = 0 \tag{3.4.1}$$

where ψ is an $s \leqslant n$ vector function. It is assumed that t_f is given explicitly. The case where t_f is given implicitly is discussed later.

Assume, for argument sake, that $u_j = u_j{}^a$ $(j = 1, \ldots, m)$ for the whole time interval, and further, that Equation (3.4.1) is satisfied using this control. In Section 3.2.1 we asserted that $V(x; t)$ has continuous partial derivatives,

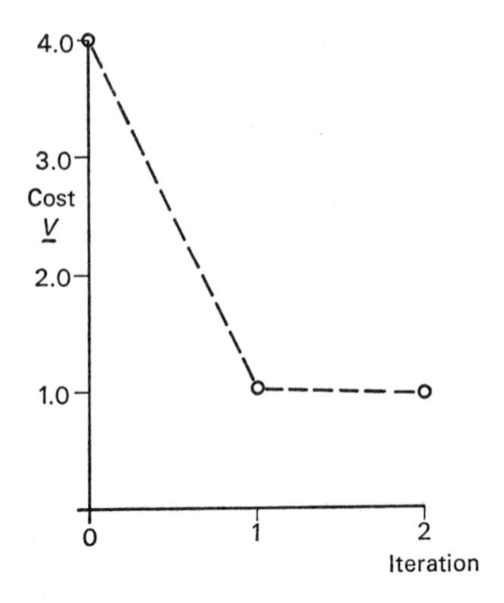

Figure 3.3. Fourth-order system: cost versus iteration number.

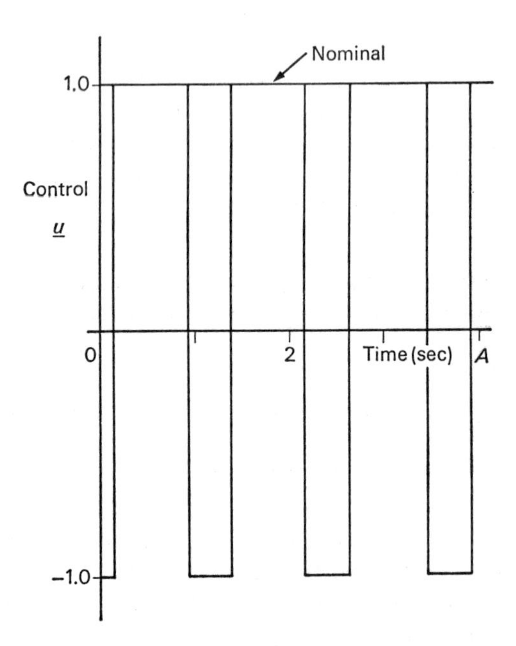

Figure 3.4. Fourth-order system: control functions.

for this constant control, in the absence of constraints (3.4.1). Now, however, these constraints are present; so there may be variations δx at time t, say, that are not allowed because they cause violation of these constraints. This is the same as saying that the partial derivatives of V with respect to x may not be defined everywhere in the state space, even for the case where the control is held constant and there are no switchings.

We may, however, convert this constrained problem into a free endpoint problem by adjoining (3.4.1) to the cost functional, using a vector Lagrange multiplier b of dimension s:

$$V(x, b; t_o) = \int_{t_o}^{t_f} L(x; t)\, dt + F(x(t_f); t_f) + \langle b, \psi(x(t_f); t_f)\rangle \tag{3.4.2}$$

For a nominal $b = \bar{b}$, this is a free endpoint problem; we assume that it has a solution.

Consider V to be a function of the multipliers b and assume a second-order expansion for V about the nominal $\bar{x}(t)$ trajectory and the nominal multipliers $\bar{b}$:

$$V(\bar{x}+\delta x, \bar{b}+\delta b; t) = \bar{V} + a + \langle V_x, \delta x\rangle + \langle V_b, \delta b\rangle + \langle \delta x, V_{xb}\delta b\rangle$$
$$+ \tfrac{1}{2}\langle \delta b, V_{bb}\delta b\rangle + \tfrac{1}{2}\langle \delta x, V_{xx}\delta x\rangle \tag{3.4.3}$$

Substituting Equation (3.4.3) into the Bellman PDE, Equation (1.4.6), and carrying out a similar derivation to that of Section 2.3.1, we obtain the following equations for the parameters of V:

$$\begin{aligned}
-\dot{a} &= H - H(\bar{x}, \bar{u}, V_x; t) \\
-\dot{V}_x &= H_x + V_{xx}(f - f(\bar{x}, \bar{u}; t)) \\
-\dot{V}_{xx} &= H_{xx} + f_x^T V_{xx} + V_{xx} f_x \\
-\dot{V}_b &= V_{xb}^T(f - f(\bar{x}, \bar{u}; t)) \\
-\dot{V}_{xb} &= f_x^T V_{xb} \\
-\dot{V}_{bb} &= 0
\end{aligned} \tag{3.4.4}$$

and at $t = t_f$, from Equation (3.4.2),

$$
\begin{aligned}
a(t_f) &= 0 \\
V_x(t_f) &= F_x(\bar{x}(t_f); t_f) + \psi_x^T(\bar{x}(t_f); t_f)\bar{b} \\
V_{xx}(t_f) &= F_{xx}(\bar{x}(t_f); t_f) + \bar{b}\psi_{xx}(\bar{x}(t_f); t_f) \\
V_b(t_f) &= \psi(\bar{x}(t_f); t_f) \\
V_{xb}(t_f) &= \psi_x^T(\bar{x}(t_f); t_f) \\
V_{bb}(t_f) &= 0
\end{aligned}
\tag{3.4.5}
$$

3.4.2. Conditions at Switch Points of the Control Function

Following the same approach as that of Section 3.3.2, the following equation in the vicinity of a switch point results:

$$
\begin{aligned}
V^-(\bar{x}+\delta x, \bar{b}+\delta b, t_s+\delta t_s; t_s) = &\int_{t_s}^{t_s+\delta t} L(\bar{x}+\delta x+\Delta x(\tau); \tau)\, d\tau \\
&+ V^+(\bar{x}+\delta x+\Delta x(t_s+\delta t_s), \bar{b}+\delta b; t_s+\delta t_s)
\end{aligned}
\tag{3.4.6}
$$

Expanding both sides of Equation (3.4.6) to second-order, and equating coefficients of like powers of δx, δb, and δt_s, we obtain the following relationships:†

$$
\begin{aligned}
\hat{V}(\bar{x}, \bar{b}; t_s) &= V(\bar{x}, \bar{b}; t_s) \\
\hat{V}_x &= V_x \\
\hat{V}_b &= V_b \\
\hat{V}_{xb} &= V_{xb} - V_{xt_s} \cdot V_{t_s b}/V_{t_s t_s} \\
\hat{V}_{xx} &= V_{xx} - V_{xt_s} \cdot V_{t_s x}/V_{t_s t_s} \\
\hat{V}_{bb} &= V_{bb} - V_{bt_s} \cdot V_{t_s b}/V_{t_s t_s}
\end{aligned}
\tag{3.4.7}
$$

† c.f. Section 3.3.2.

where

$$V_{bt_s} = V_{xb}^T(f^- - f^+) \tag{3.4.8}$$

and V_{xt_s} and $V_{t_st_s}$ are given by Equation (3.3.24).

The following local, linear controller relates t_s to b and $x(t_s)$:

$$\delta t_s = -V_{t_st_s}^{-1}[\langle V_{bt_s}, \delta b\rangle + \langle V_{xt_s}, \delta x\rangle] \tag{3.4.9}$$ †

3.4.3. The Computational Procedure

First b in Sections 3.4.1-3.4.2 is set to a nominal value of $\bar{b}$. The $\dot{a}$, $\dot{V}_x$ and $\dot{V}_{xx}$ equations are integrated backward, and jumps in V_{xx} are calculated at switch times. The computational procedure of Section 3.3.5 is used to solve this free endpoint problem. Along the optimal trajectory the following equations hold:

$$\begin{aligned} -\dot{a} &= 0 \\ -\dot{V}_x &= H_x \\ -\dot{V}_b &= 0 \\ -\dot{V}_{xb} &= f_x^T V_{xb} \\ -\dot{V}_{bb} &= 0 \\ -\dot{V}_{xx} &= H_{xx} + f_x^T V_{xx} + V_{xx} f_x \end{aligned} \tag{3.4.10}$$

The differential equation for V_{xb} may now be integrated backward along this trajectory. At switch points, Equations (3.4.7) determine the jumps in V_{xb} and V_{bb}. Notice that V_{bb} is a piecewise constant function of time.

At $t = t_o$,

$$V(x_o, \bar{b}+\delta b; t_o) = V(x_o, \bar{b}; t_o) + \langle V_b, \delta b\rangle + \tfrac{1}{2}\langle \delta b, V_{bb}\delta b\rangle \tag{3.4.11}$$

From Equations (3.4.9) and (3.4.10),

$$V_b(x_o, \bar{b}; t_o) = \psi(\bar{x}(t_f); t_f) \tag{3.4.12}$$

† Equation (3.4.9) is obtained in an exactly analogous fashion to Equation (3.3.31).

In order to reduce $V_b(t_o)$, and hence ψ, to zero, we differentiate Equation (3.4.11) with respect to δb and equate to zero:

$$V_b\Big|_{t_o} + V_{bb}\Big|_{t_o} \delta b = 0 \tag{3.4.13}$$

whence

$$\delta b = -\varepsilon V_{bb}^{-1} V_b\Big|_{t_o} \tag{3.4.14}$$

where ε $(0 < \varepsilon \leqslant 1)$ is present to ensure that δb is not so large that it invalidates the above expansions. For ε sufficiently small, Equation (3.4.14) ensures a reduction in V_b.

δb, given by Equation (3.4.14) is then used in Equation (3.4.1) to compute the required changes in switch times that will result in a reduction in the endpoint error ψ.

Equation (3.4.7) becomes

$$\delta t_{s_i} = -V_{t_s t_s}^{-1}(\bar{x}, \bar{b}, t_{s_i}; t_{s_i})[\langle V_{bt_s}(\bar{x}, \bar{b}, t_{s_i}; t_{s_i}), -\varepsilon V_{bb}^{-1} V_b\Big|_{t_o}\rangle$$
$$+ \langle V_{xt_s}(\bar{x}, \bar{b}, t_{s_i}; t_{s_i}), \delta x(t_{s_i})\rangle] \tag{3.4.15}$$

The computational procedure summarized:

1. Set $b = \bar{b}$ and solve the resulting free endpoint problem using the method of Section 3.3.5.
2. Integrate the $\dot{V}_{xb}$ equation backward and compute the jump in V_{xb} and V_{bb} at switch times of $u^*(t)$. Use Equation (3.4.15) to compute the new control function.† For ε sufficiently small‡ there will be an improvement in the terminal condition error. Check $|a(x_o; t_o)|$; if this is less than η_1, repeat Step 2 to reduce the terminal error further; if $|a(x_o; t_o)|$ is greater than η_1, repeat Step 1.

When $|a(x_o; t_o)| < \eta_1$ and $|V_b(x_o, b; t_o)| < \eta_2$, stop the computation ($\eta_1$ and η_2 are small, positive quantities).

3.4.4. Characteristics of the Algorithm

In addition to the points listed in Section 3.3.6, the following is important: in order for the matrix $V_{bb}(x_o, b; t_o)$ to be invertible, there must be at least s switchings of the control u^* in the interval $[t_o, t_f]$ (b is s-dimensional).

---

† Using the procedure of Section 3.3.4.

‡ ε is chosen experimentally.

It is possible that the guessed multipliers $\bar{b}$, and nominal trajectory $\bar{x}(t)$; $t \in [t_o, t_f]$, might be so far from optimal that the resulting u^* does not have s switchings. In this case, δb may be calculated by using the first-order expression $\delta b = \varepsilon V_b$, or $\delta b = -\varepsilon V_{bb}^{+} V_b$, where V_{bb}^{+} is the pseudo-inverse of V_{bb}.

3.4.5. Sufficient Conditions for an Improved Trajectory at Each Iteration

Sufficient conditions to guarantee $a(x_o, \bar{b}; t_o) < 0$, and hence a reduction in cost for the free endpoint problem with $b = \bar{b}$, have been given in Section 3.3.7.

A sufficient condition for a reduction in terminal error is, from Equation (3.4.14),

$$V_{bb}(x_o, \bar{b}; t_o) \qquad \text{be invertible} \tag{3.4.16}$$

Sufficient conditions to ensure (3.4.16) are:

1. There must be at least s† switchings of the control $u^*(t)$ in the interval $[t_o, t_f]$; i.e., $n_s \geqslant s$.
2. The vectors $V_{bt_s}(\bar{x}, \bar{b}, t_{s_j}; t_{s_j})$, s of them, must be linearly independent $j \in \{1, \ldots, n_s\}$.

PROOF: From Equations (3.4.4), (3.4.5) and (3.4.7),

$$V_{bb}(x_o, \bar{b}; t_o) = \sum_{i=1}^{n_s} \left\{ \frac{V_{bt_s}(\bar{x}, \bar{b}, t_{s_i}; t_{s_i}) \cdot V_{t_s b}(\bar{x}, \bar{b}, t_{s_i}; t_{s_i})}{V_{t_s t_s}(\bar{x}, \bar{b}, t_{s_i}; t_{s_i})} \right\} \tag{3.4.17}$$

Because $V_{t_s t_s}$ is assumed positive (Section 3.3.2) and the dyads $V_{bt_s} \cdot V_{t_s b}$ are positive-semidefinite, V_{bb} is at worst negative-semidefinite. If V_{bb} is negative-definite, then it will surely be invertible. Clearly, for V_{bb} to be negative-definite, it must have full rank s; thus in the above summation, n_s must be greater or equal to s, and s of the dyads must be linearly independent.

3.4.6. Computed Example

Using the same dynamics as Example II, Section 3.3.9, we wish to minimize

$$V = x_4(t_f) \tag{3.4.18}$$

† b is s-dimensional.

subject to

$$x_1(t_f) = 2.3, \qquad x_2(t_f) = 2.4, \qquad x_3(t_f) = 1.5 \tag{3.4.19}$$

where t_f is set as 2.5 sec.

The following nominal values are used:

$$\begin{aligned} \bar{u}(t) &= +1; \qquad t \in [0, 2.5] \\ \bar{b}_1 &= \bar{b}_2 = \bar{b}_3 = 0.50 \end{aligned} \tag{3.4.20}$$

Using the above nominal control, it was found that

$$\begin{aligned} x_1(t_f) = 2.81; \qquad x_2(t_f) &= 3.06; \qquad x_3(t_f) = 1.99 \\ x_4(t_f) &= 2.51 \end{aligned} \tag{3.4.21}$$

After seven iterations of the procedure, the following data was obtained:

$$\begin{aligned} x_1(t_f) = 2.30; \qquad x_2(t_f) &= 2.40; \qquad x_3(t_f) = 1.50 \\ x_4(t_f) &= 2.31 \end{aligned} \tag{3.4.22}$$

The optimal values of the Lagrange multipliers are

$$b_1 = 1.07; \qquad b_2 = 0.98; \quad b_3 = 0.99 \tag{3.4.23}$$

The optimal control has six switchings:

$$n_s = 6 > s = 3 \tag{3.4.24}$$

and is similar in form to that shown in Figure 3.4.

3.4.7. Final Time Given Implicitly

Here t_f is treated in the same way as in Section 2.3.5; i.e., t_f is imbedded in V:

$$V(x_o, b, t_f; t_o) = \int_{t_o}^{t_f} L(x; t)\, dt + F(x(t_f); t_f) + \langle b, \psi(x(t_f); t_f) \rangle \tag{3.4.25}$$

In a similar way to that demonstrated in previous sections, differential equations and jump conditions can be obtained for V_{t_f}, V_{xt_f}, V_{bt_f}, and $V_{t_f t_f}$.

A relationship of the form

$$\delta t_s = -V_{t_s t_s}^{-1}[V_{t_f t_s}\delta t_f + \langle V_{bt_s}, \delta b\rangle + \langle V_{xt_s}, \delta x\rangle] \qquad (3.4.26)$$

relates the changes in switch times to changes in x, k, and t_f.

3.5. DYNAMIC PROGRAMMING AND PONTRYAGIN'S MINIMUM PRINCIPLE

3.5.1. An Interpretation for Pontryagin's Adjoint Variables in Bang-Bang Control Problems

In this section we draw certain comparisons between DP and Pontryagin's principle; in particular, we give an interpretation for Pontryagin's adjoint variables λ, for the class of problems treated in this chapter.

For the free-endpoint problem, it is clear that $\lambda = V_x{}^{\text{o}}$ because $V_x{}^{\text{o}}$ has been shown to be continuous and to satisfy the same differential equation as λ. The case of fixed-endpoint problems is not as simple. Consider the cost functional:‡

$$V(x_o, b; t_o) = \int_{t_o}^{t_f} L(x; t)\, dt + F(x(t_f); t_f) + \langle b, \psi(x(t_f); t_f)\rangle \qquad (3.5.1)$$

Equation (3.5.1) describes the optimal cost for the free-endpoint problem obtained by afdoining ψ to Equation (3.2.3); i.e., V is minimized and b is chosen such that

$$\psi(x(t_f); t_f) = 0 \qquad (3.5.2)$$

Consider also the following cost functional:

$$V^{\text{o}}(x_o; t_o) = \int_{t_o}^{t_f} L(x; t)\, dt + F(x(t_f); t_f) \qquad (3.5.3)$$

$$\psi(x(t_f); t_f) = 0 \qquad (3.5.4)$$

---

† Equation (3.4.26) is obtained in an exactly analogous fashion to Equation (3.3.31).

‡ In this section optimal control and state values are denoted by u and x, with no superscripts or "bars."

Equation (3.5.3) describes the optimal cost† for the fixed-endpoint problem obtained without adjoining (3.5.4). Both in Equations (3.5.1) and (3.5.3), t_f is assumed to be given explicitly.

Dynamic programming in its continuous form requires the optimal cost surfaces $V(x, b; t)$ and $V^\circ(x; t)$ to have continuous partial derivatives with respect to x, b, and t, for the derivation of the Bellman PDE to be valid. In this chapter we have shown that $V(x, b; t)$ has continuous partial derivatives except at switch points of the control function, where special jump conditions have been investigated. It is well-known that in problems with bang-bang control and fixed-endpoint, the optimal cost surface V° is nonsmooth across some switching surfaces in the state space. In fact V_x°, the first partial derivative of V° with respect to x, is not continuous across some switching surfaces.‡ It so happens that the optimal trajectory never crosses these surfaces (where V_x° is not defined) but travels along them [8].

It is the purpose of this section to obtain an interpretation for Pontryagin's adjoint variable λ for this class of problems, where the system trajectory is tangent to the manifold of discontinuous control.

Berkovitz [9] has related Pontryagin's principle to dynamic programming, but does not consider this "tangency case."

One is able to convert the free-endpoint formulation, Equation (3.5.1), into the fixed-endpoint formulation, Equations (3.5.3) and (3.5.4), locally, in the following way: in the neighborhood of an optimal trajectory, one has, to second-order, that

$$V(x+\delta x, b+\delta b; t) = V(x, b; t) + \langle V_x, \delta x\rangle + \langle V_b, \delta b\rangle + \langle \delta x, V_{xb}\delta b\rangle$$
$$+ \tfrac{1}{2}\langle \delta b, V_{bb}\delta b\rangle + \tfrac{1}{2}\langle \delta x, V_{xx}\delta x\rangle \qquad (3.55.)$$

Because the trajectory is optimal,

$$V_b(x, b; t) = \psi(x(t_f); t_f) = 0 \qquad (3.5.6)$$

Now let us eliminate δb from Equation (3.5.5) in such a way that the endpoint condition is still satisfied; i.e., determine δb such that

$$\psi(x(t_f) + \delta x(t_f); t_f) = 0 \qquad (3.5.7)$$

† From Equation (3.5.3) onward, the superscript $^\circ$ on V denotes optimal cost for the problem with fixed-endpoint. No superscript denotes optimal cost for the problem where the end constraints have been adjoined to the cost functional by Lagrange multipliers b.

‡ The works by Dreyfus [8], Knudsen [10], Fuller [11], Kalman [12], and Shapiro [13] have direct relevance.

From Equations (3.5.5), (3.5.6), and (3.5.7),

$$V_b(x+\delta x, b+\delta b; t) = V_{xb}^T \delta x + V_{bb} \delta b = 0 \tag{3.5.8}$$

Whence

$$\delta b = -V_{bb}^{-1} V_{xb}^T \delta x \tag{3.5.9}$$

Substituting Equation (3.5.9) into Equation (3.5.5),

$$V(x+\delta x, b - V_{bb}^{-1} V_{xb}^T \delta x; t)$$
$$= V + \langle V_x, \delta x \rangle + \tfrac{1}{2} \langle \delta x, (V_{xx} - V_{xb} V_{bb}^{-1} V_{bx}) \delta x \rangle \tag{3.5.10}$$

This is an expression for the optimal cost, locally, for the fixed-endpoint formulation; so in in view of Equations (3.5.3) and (3.5.4), which represent just this problem, we may write Equation (3.5.10) as

$$V^{\circ}(x+\delta x; t) = V^{\circ} + \langle V_x^{\circ}, \delta x \rangle + \tfrac{1}{2} \langle \delta x, (V_{xx} - V_{xb} V_{bb}^{-1} V_{bx}) \delta x \rangle \tag{3.5.11}$$

Notice that V_x°, for this optimal cost function, is the samc as the V_x, for the optimal free-endpoint problem, but that

$$V_{xx}^{\circ} = V_{xx} - V_{xb} V_{bb}^{-1} V_{bx} \tag{3.5.12}$$

However, Equation (3.5.12) holds only if V_{bb} is invertible; so the transformation (3.5.9) cannot be done unless V_{bb} is invertible.† This means that one cannot convert the free-endpoint formulation, Equation (3.5.1), into the fixed-endpoint formulation, Equations (3.5.3) and (3.5.4), in regions of the state space where V_{bb} is not invertible; thus V_x° cannot be identified with V_x. Since the equations for Pontryagin's λ are exactly the same as those for V_x for the free-endpoint problem (and also the same for V_x° when Equations (3.5.9) and (3.5.10) are valıd) one has the following simple interpretation for λ: in regions where V_{bb} is invertible, $\lambda = V_x^{\circ} = V_x$. In regions where V_{bb} is not invertible, $\lambda = V_x$.

The above section does, it is believed, help to close the "conceptual gap" between DP and Pontryagin's principle. The difference in approach between the above study and that of others (notably Shapiro [13]) is that λ is identified with a quantity V_x that is the partial derivative of the optimal cost for the equivalent free-endpoint problem. The usual approach is to try and identify λ

† Conditions for this are given in Section 3.4.5. It is sufficient to note that V_{bb} can be invertible only after, at least, s switchings of the control function. This is because $\dot{V}_{bb} = 0$ and V_{bb} experiences a jump at switch points of the control function.

with a quantity associated with the optimal cost function V^o; this is difficult. Dreyfus [8]†, when considering the case of trajectories tangent to the switching manifold, points out that the classical multipliers cannot be interpreted as partial derivatives of the optimal value function. He states further that the classical multipliers apparently have no physical or geometrical interpretation. It has been shown in this section that λ does have a simple physical and geometrical interpretation.

3.6. SUMMARY

This chapter has been concerned with the rather special class of control problems whose optimal controls are piecewise constant functions of time. The novel approach of differential dynamic programming has yielded new insight into this class of control problems. In particular we have shown that, for the free-endpoint problem, the first partial derivatives of the optimal cost function are continuous throughout the state space and the second partial derivatives experience jumps at switch points of the control function. Inequality (3.5.16) is a new necessary condition of optimality for bang-bang control problems. New second-order and first-order algorithms were developed using the above results.

In Section 3.5 we gave an interpretation for Pontryagin's adjoint variables for this class of problems. This clarifies some previous attempts at bridging the gap between Pontryagin's principle and dynamic programming [8, 12, 13].

References

1. H. J. Kelley, R. E. Kopp, and H. Gardner Moyer, *in* "Topics in Optimization" (G. Leitman, ed.), Chapter 3. Academic Press, New York, 1967.
2. H. M. Robbins, IBM Rept. No. 66-825 2043 (1966).
3. K. S. Tait, Ph.D. Thesis, Harvard Univ., Cambridge, Massachusetts, 1965.
4. C. D. Johnson, *in* "Advances in Control Systems" (C. T. Leondes, ed.), Vol. 2. Academic Press, New York, 1965.
5. D. H. Jacobson, *IEEE Trans. Auto. Control,* **AC-13**, 661 (1968).
6. P. Dyer and S. R. McReynolds, *J. Math. Anal. Appl.*, **23**, 525 (1968) in press.
7. J. B. Plant and M. Athans, *Proc. 3rd Congr. Intern. Fed. Auto. Control, London, 1966.*
8. S. E. Dreyfus, "Dynamic Programming and the Calculus of Variations," p. 190. Academic Press, New York, 1965.
9. L. D. Berkovitz, *J. Math. Anal. Appl.* **3**. 145 (1961).
10. H. K. Knudsen, *IEEE Trans. Auto. Control* **AC-9**, 23 (1964).
11. A. T. Fuller, *J. Electron. Control* **17**, 283 (1964).
12. R. E. Kalman, *in* "Mathematical Optimization Techniques" (R. Bellman, ed.), California Univ. Press, Berkeley, California, 1963.
13. S. Shapiro, Ph.D. Thesis, Univ. of London, England, 1965.
14. B. Friedlaud and P.E. Sarachik, Proc. 3rd Congr. Intern. Fed. Auto. Control, London, 1966.

† See his chapter on problems with special linear structures, p. 204.

Chapter 4

DISCRETE-TIME SYSTEMS

4.1 INTRODUCTION

In this chapter we shall consider systems described by nonlinear difference equations. In one respect discrete-time systems are simpler to analyze than the continuous-time systems considered in the first three chapters—differential equations are replaced by difference equations whose solutions are easier to compute. However discrete-time systems also produce complications of their own. A noninfinitesimal change in the control action u_i at time i produces noninfinitesimal changes in the subsequent trajectory. For the continuous-time system, on the other hand, a noninfinitesimal change in the control action $u(t)$ over the interval $[t_1 - \varepsilon, t_1 + \varepsilon]$ ($\varepsilon > 0$ but arbitrarily small) produces small, or order ε, variations in $x(t)$ $(t > t_1)$. It is this property of continuous-time systems that permits the derivation of a global minimum principle, and the derivation of algorithms, described in Chapters 2 and 3, employing global variations in control. For discrete-time systems, a global minimum principle holds only under very restrictive conditions, and the derivation of optimization algorithms requires extra care.

The system that we shall consider is defined by

$$x_{i+1} = f_i(x_i, u_i), \qquad i = 0, 1, \ldots, N-1 \tag{4.1.1}$$

$$x_0 = \bar{x}_0 \tag{4.1.2}$$

where x_i is the n-dimensional state vector, and u_i is the m-dimensional control vector, the suffix i denoting time i. f_i is an n-dimensional vector function of (x_i, u_i). The $n \times n$ matrix whose rsth component is $\partial (f_i)_r / \partial (x_i)_s$ will be denoted $f_x^{\,i}$ [$(\;)_r$ denotes the rth component of a vector and $(\;)_{rs}$ the rsth component of a matrix]. A sequence of control actions $\{u_i, i = 0, \ldots, N-1\}$ will often be denoted by $\{u_i\}$ and referred to as a schedule or control trajectory. The sequence of states $\{x_i, i = 0, \ldots, N\}$ will be denoted by $\{x_i\}$ and referred to as a state trajectory.

The cost function, or performance index, of a trajectory with initial condition x_o and schedule $\{u_i\}$ is

$$V_o(x_o, \{u_i\}) = \sum_{i=0}^{N-1} L_i(x_i, u_i) + F(x_N) \tag{4.1.3}$$

where L_i is a nonnegative function of (x_i, u_i); F is a nonnegative function of (x_N); and $x_o, x_1, \ldots, x_N$ satisfy Equation (4.1.1) for the given schedule $\{u_i\}$. Hence V_o is a function of x_o, $\{u_i\}$.

The unconstrained problem that shall be our main concern is to find the sequence $\{u_i\}$ that minimizes $V_o(\bar{x}_o, \{u_i\})$. The algorithms presented in this chapter are easily modified to deal with terminal constraints of the form:

$$\psi(x_N) = 0 \tag{4.1.4}$$

where ψ is an $s \leqslant n$-dimensional vector function of x_N. The constraints may be incorporated using the technique described in Section 2.3. Control constraints will not be considered.

A nominal control sequence is denoted by $\{u_i\}$. The corresponding state sequence [the sequence of states satisfying Equations (4.1.1) and (4.1.2) with $u_i = \bar{u}_i$, $i = 0, \ldots, N-1$] is denoted $\{\bar{x}_i\}$. The cost due to the schedule $\{\bar{u}_i\}$ and initial condition x_o is denoted $\bar{V}_o(x_o)$. Clearly,

$$\bar{V}_o(x_o) = V_o(x_o, \{\bar{u}_i\}) \tag{4.1.5}$$

To employ dynamic programming, it is necessary to define the cost of a trajectory with initial condition x_i (at time i). This cost is

$$V_i(x_i, \{u_k\}) = \sum_{k=i}^{N-1} L_k(x_k, u_k) + F(x_N) \tag{4.1.6}$$

where $\{u_k\}$ denotes here the sequence $\{u_i, \ldots, u_{N-1}\}$; and $x_i, \ldots, x_N$ satisfy Equation (4.1.1) for this schedule. Similarly,

$$\bar{V}_i(x_i) = V_i(x_i, \{\bar{u}_k\}) \tag{4.1.7}$$

is the cost due to initial condition x_i and schedule $\{\bar{u}_k\}$.

Let $\{u_i^o\}$ denote the optimal schedule, and $\{x_i^o\}$ the resultant state trajectory for initial condition $\bar{x}_o$. Let π denote any policy (i.e., sequence of control laws $u_k = h_k(x_k)$, $k = 0, \ldots, N-1$) and π^o the optimal policy. Let $V_i^o(x_i)$ denote the cost due to initial condition x_i if the optimal policy is employed. Then, by the principle of optimality,

$$V_i^{\circ}(x_i) = \min_{u_i} [L_i(x_i, u_i) + V_{i+1}^{\circ}(f_i(x_i, u_i))] \tag{4.1.8}$$

Performing the minimization indicated in Equation (4.1.8) yields the optimal control action and the optimal cost for state x_i. Repeating the minimization for all $x_i \in E_n$ yields the functions $V_i^{\circ}(\cdot)$, the optimal cost function, and $h_i^{\circ}(\cdot)$, the optimal control law, at time i. Iteration of this procedure for $i = N-1, \ldots, 0$, using the boundary condition

$$V_N^{\circ}(x_N) = F(x_N) \tag{4.1.9}$$

yields $V_0^{\circ}(\cdot) \ldots V_N^{\circ}(\cdot)$ and also the optimal policy $\pi^{\circ} = \{h_0^{\circ}(\cdot), \ldots, h_{N-1}^{\circ}(\cdot)\}$. We thus have the information necessary to determine the optimal control sequence for any initial condition. In particular the optimal sequence $\{u_0^{\circ}, \ldots, u_{N-1}^{\circ}\}$, for the initial condition $\bar{x}_0$, is obtained by solving the following difference equations:

$$x_{i+1}^{\circ} = f_i(x_i^{\circ}, u_i^{\circ}), \qquad x_0^{\circ} = \bar{x}_0 \tag{4.1.10}$$

$$u_i^{\circ} = h_i^{\circ}(x_i^{\circ}) \tag{4.1.11}$$

Dynamic programming clearly requires the storage of $V_i^{\circ}(x_i)$, $h_i^{\circ}(x_i)$ for all $x_i \in E_n$, $i = 0, \ldots, N-1$, and this is usually impossible. Differential dynamic programming avoids this difficulty by comparing the nominal trajectory $\{\bar{x}_i\}$ with neighboring trajectories $\{x_i\}$ and selecting that neighboring trajectory which yields an optimal decrease in cost. The new control and state sequences become the nominal sequences for the next iteration of the algorithm. Since, at each iteration of the algorithm, attention is confined to a suitably small neighborhood of $\bar{x}_i$, $i = 0, \ldots, N$, it is possible to approximate the improved cost function over this neighborhood by a truncated Taylor series expansion, and all that is required to be stored are the coefficients of this Taylor series expansion. This results in a tremendous reduction in storage required compared with dynamic programming. A disadvantage is that many iterations of the algorithm are required as compared with the one iteration required by dynamic programming.

In Section 4.3 we show how differential dynamic programming may be employed to obtain an optimization algorithm using expansions about the nominal sequences $\{u_i\}$ and $\{\bar{x}_i\}$. Our main requirement is that the variations in the state from the nominal states due to the new control sequence should be sufficiently small, and this may be achieved even if the variation in control actions is large, provided the time duration of this variation is small. Hence in Section 4.4. we derive an optimization algorithm employing global variations in control. In order to pave the way for a discussion of stochastic systems in Chapters 5 and 6, we shall examine nonoptimal policies in Sections 4.3 and 4.4. The work in Section 4.3 is based on the paper by Mayne[1] and that in Section 4.4 on a report by Gershwin and Jacobson [2].

4.2. NOMINAL CONTROL SEQUENCE

In this section we show how the parameters, which define $\overline{V}_i(x_i)$ (the cost of a trajectory with initial condition x_i and schedule $\{\bar{u}_k\}$) in a sufficiently small neighborhood of $\bar{x}_i$, $i=0, \ldots, N$, may be calculated. Let $\overline{V}_x{}^i(\bar{x}_i)$ denote the column vector whose rth component is $\partial/\partial(x_i)_r[V_i(\bar{x}_i)]$, $r=1, \ldots, n$; and let $\overline{V}_{xx}^i(\bar{x}_i)$ denote the $n \times n$ symmetric matrix whose rsth component is $\partial^2/\partial(x_i)_r \partial(x_i)_s[\overline{V}_i(\bar{x}_i)]$, r, $s=1, \ldots, n$. Then, if $\overline{V}_i(x_i)$ is a sufficiently smooth function of x_i, it can be approximated in a sufficiently small neighborhood of $\bar{x}_i$ by the following truncated Taylor series:

$$\overline{V}_i(x_i) = \overline{V}_i(\bar{x}_i) + [\overline{V}_x{}^i(\bar{x}_i)]^T \delta x_i + \tfrac{1}{2} \delta x_i{}^T [\overline{V}_{xx}^i(\bar{x}_i)] \delta x_i + \cdots \tag{4.2.1}$$

If the third derivative of $\overline{V}_i(x_i)$ exists, and is bounded, then the error in Equation (4.2.1) is of third-order in δx_i. The parameters $\overline{V}_i(\bar{x}_i)$, $\overline{V}_x{}^i(\bar{x}_i)$, $\overline{V}_{xx}^i(\bar{x}_i)$ are the parameters that we wish to calculate.

A feature of dynamic programming (and of differential dynamic programming) that has not been emphasized previously is that it can be used to determine the cost of nonoptimal control sequences or policies. Thus, from Equations (4.1.6) and (4.1.7), we have

$$\overline{V}_i(x_i) = L_i(x_i, \bar{u}_i) + \overline{V}_{i+1}(x_{i+1}) \tag{4.2.2}$$

where

$$x_{i+1} = f_i(x_i, \bar{u}_i) \tag{4.2.3}$$

and

$$\overline{V}_N(x_N) = F(x_N) \tag{4.2.4}$$

Equations (4.2.2) to (4.2.4) define an algorithm for the successive determination of the functions $\overline{V}_N(\cdot), \ldots, \overline{V}_0(\cdot)$. If, however, we only require knowledge of $\overline{V}_i(x_i)$ in the neighborhood of $\bar{x}_i$, then the parameters $\overline{V}_i(\bar{x}_i)$, $\overline{V}_x{}^i(\bar{x}_i)$, $\overline{V}_{xx}^i(\bar{x}_i)$ that yield an approximation of $\overline{V}_i(x_i)$ in this neighborhood may be obtained by a procedure that requires much less storage and computation.

From Equation (4.2.2), setting $x_i = \bar{x}_i$ and, hence, $x_{i+1} = \bar{x}_{i+1}$, we obtain

$$\bar{V}_i(\bar{x}_i) = L_i(\bar{x}_i, \bar{u}_i) + \bar{V}_{i+1}(\bar{x}_{i+1}) \tag{4.2.5}$$

Differentiating both sides of Equation (4.2.2) with respect to x_i yields

$$\bar{V}_x{}^i(x_i) = L_x{}^i(x_i, \bar{u}_i) + [f_x{}^i(x_i, \bar{u}_i)]^T \bar{V}_x^{i+1}(x_{i+1}) \tag{4.2.6}$$

so that

$$\bar{V}_x{}^i(\bar{x}_i) = L_x{}^i(\bar{x}_i, \bar{u}_i) + [f_x{}^i(\bar{x}_i, \bar{u}_i)]^T \bar{V}_x^{i+1}(\bar{x}_{i+1}) \tag{4.2.7}$$

To obtain an iterative equation for $\bar{V}_{xx}^i(\bar{x}_i)$ it is necessary to differentiate Equation (4.2.6) with respect to x_i, which causes notational difficulties.It is useful, therefore, to introduce a pseudo-Hamiltonian function defined by

$$H_i(x_i, u_i, \lambda) = L_i(x_i, u_i) + \lambda^T f_i(x_i, u_i), \qquad i = 0, \ldots, N-1. \tag{4.2.8}$$

In terms of H, Equations (4.2.6) and (4.2.7) become

$$\bar{V}_x{}^i(x_i) = H_x{}^i(x_i, \bar{u}_i, \bar{V}_x^{i+1}(x_{i+1})) \tag{4.2.9}$$

$$\bar{V}_x{}^i(\bar{x}_i) = H_x{}^i(\bar{x}_i, \bar{u}_i, \bar{V}_x^{i+1}(\bar{x}_{i+1})) \tag{4.2.10}$$

Differentiating Equation (4.2.9) with respect to x_i, and setting $x_i = \bar{x}_i$, yields:

$$\bar{V}_{xx}^i(\bar{x}_i) = H_{xx}^i(\bar{x}_i, \bar{u}_i, \bar{V}_x^{i+1}(\bar{x}_{i+1})) + [f_x{}^i(\bar{x}_i, \bar{u}_i)]^T [\bar{V}_{xx}^{i+1}(\bar{x}_{i+1})][f_x{}^i(\bar{x}_i, \bar{u}_i)] \tag{4.2.11}$$

Summarizing, the difference equations for $\bar{V}_i(\bar{x}_i)$, $\bar{V}_x{}^i(\bar{x}_i)$ and $\bar{V}_{xx}^i(\bar{x}_i)$ are

$$\begin{aligned}
\bar{V}_i(\bar{x}_i) &= L_i(\bar{x}_i, \bar{u}_i) + \bar{V}_{i+1}(\bar{x}_{i+1}) \\
\bar{V}_x{}^i(\bar{x}_i) &= H_x{}^i(\bar{x}_i, \bar{u}_i, \bar{V}_x^{i+1}(\bar{x}_{i+1})) \\
\bar{V}_{xx}^i(\bar{x}_i) &= H_{xx}^i(\bar{x}_i, \bar{u}_i, \bar{V}_x^{i+1}(\bar{x}_{i+1})) \\
&\quad + [f_x{}^i(\bar{x}_i, \bar{u}_i)]^T [\bar{V}_{xx}^{i+1}(\bar{x}_{i+1})][f_x{}^i(\bar{x}_i, \bar{u}_i)]
\end{aligned} \tag{4.2.12}$$

for $i=0, \ldots, N-1$. The boundary conditions for Equations (4.1.23) are obtained from Equation (4.2.4):

$$\bar{V}_N(\bar{x}_N) = F(\bar{x}_N)$$

$$\bar{V}_x^N(\bar{x}_N) = F_x(\bar{x}_N) \tag{4.2.13}$$

$$\bar{V}_{xx}^N(\bar{x}_N) = F_{xx}(\bar{x}_N)$$

Using Equations (4.2.12) and (4.2.13), $\bar{V}_i(\bar{x}_i)$, $\bar{V}_x{}^i(\bar{x}_i)$ and $\bar{V}_{xx}^i(\bar{x}_i)$, $i=0, \ldots, N$, can be calculated in reverse time, yielding the desired Taylor series approximation to $\bar{V}_i(x_i)$ in a suitably small neighborhood of $\bar{x}_i$.

Equations (4.2.12) and (4.2.13) are merely a disguised form of the "chain rule" for differentiation. This is easily seen for the case when $L_i=0$, $i=0, \ldots, N-1$, when repeated iteration of Equation (4.2.6) yields

$$\bar{V}_x{}^{\mathrm{o}}(\bar{x}_{\mathrm{o}}) = [f_x{}^{\mathrm{o}}]^T, \ldots, [f_x^{N-1}]^T F_x(\bar{x}_N)$$

the unspecified arguments of $f_x{}^i$ being $\bar{x}_i$, $\bar{u}_i$.

The above discussion can be trivially extended to obtain the coefficients of the Taylor series expansion of $V_i(x_i, \pi)$ about $\bar{x}_i$, where $V_i(x_i, \pi)$ is the cost due to initial condition x_i and policy $\pi = \{h_{\mathrm{o}}(\cdot), \ldots, h_{N-1}(\cdot)\}$, where

$$u_i = h_i(x_i), \qquad i=0, \ldots, N-1 \tag{4.2.14}$$

For convenience, define f_i', L_i', H_i' as follows:

$$[f_i(x_i)]' = f_i(x_i, h_i(x_i)) \tag{4.2.15}$$

$$[L_i(x_i)]' = L_i(x_i, h_i(x_i)) \tag{4.2.16}$$

$$[H_i(x_i, \lambda)]' = H_i(x_i, h_i(x_i), \lambda) \tag{4.2.17}$$

Suppose that the nominal state and control sequences have been generated as follows:

$$\bar{x}_{i+1} = f_i(\bar{x}_i, \bar{u}_i), \qquad x_{\mathrm{o}} = \bar{x}_{\mathrm{o}} \tag{4.2.18}$$

$$\bar{u}_i = h_i(\bar{x}_i) \tag{4.2.19}$$

i.e.,

$$\bar{x}_{i+1} = [f_i(\bar{x}_i)]' \tag{4.2.20}$$

for $i=0, \ldots, N-1$. Then,

$$\begin{aligned} V_i(\bar{x}_i, \pi) &= [L_i(\bar{x}_i)]' + V_{i+1}(\bar{x}_{i+1}, \pi) \\ V_x{}^i(\bar{x}_i, \pi) &= [H_x{}^i(\bar{x}_{i+1}, V_x^{i+1}(\bar{x}_{i+1}, \pi)]' \\ V_{xx}^i(\bar{x}_i, \pi) &= [H_{xx}^i(\bar{x}_{i+1}, V_x^{i+1}(\bar{x}_{i+1}, \pi)]' \\ &\quad + [(f_x{}^i(\bar{x}_i))']^T [V_{xx}^{i+1}(\bar{x}_{i+1}, \pi)][(f_x{}^i(\bar{x}_i))'] \end{aligned} \tag{4.2.21}$$

for $i = 0, \ldots, N-1$. The boundary conditions are specified by Equation (4.2.13), $V_N(\bar{x}_N, \pi)$ replacing $\bar{V}_N(\bar{x}_N)$, etc.

4.3. DIFFERENTIAL DYNAMIC PROGRAMMING—SMALL VARIATIONS IN CONTROL

4.3.1. Local Linear Nonoptimal Control Policy

Suppose that the nominal sequences $\{\bar{u}_k\}$ and $\{\bar{x}_k\}$ are known, and consider the problem of estimating $V_i(x_i, \pi)$ in the neighborhood of $\bar{x}_i, i=0, \ldots, N$, where the policy $\pi = \{h_0(\cdot), \ldots, h_{N-1}(\cdot)\}$ is defined by

$$u_i = \bar{u}_i + \delta u_i \tag{4.3.1}$$

$$\delta u_i = \alpha_i + \beta_i \delta x_i \tag{4.3.2}$$

$$\delta x_i = x_i - \bar{x}_i \tag{4.3.3}$$

for $i=0, \ldots, N-1$, where α_i, $i=0, \ldots, N-1$, is sufficiently small to justify certain approximations discussed later. Application of the above policy (using initial condition $\bar{x}_0$) yields the sequences $\{u_k\}$ and $\{x_k\}$. The magnitude of $\delta x_i, i=0, \ldots\ N$, is controlled by the magnitude of $\alpha_i, i=0, \ldots, N-1$. For, if $\alpha_i = 0$, $i = 0, \ldots, N-1$, then $\delta x_i = 0$, $i = 0, \ldots, N$, since $\delta x_0 = 0\ (x_0 = \bar{x}_0)$.

We wish to obtain a procedure for calculating the Taylor series coefficients $V_i(\bar{x}_i)$, $V_x^i(\bar{x}_i)$, $V_{xx}^i(\bar{x}_i)$, $i=0, \ldots, N$; the argument π is omitted for notational simplicity. Clearly $V_i(x_i)$ satisfies the following differences equation:

$$V_i(x_i) = L_i(x_i, u_i) + V_{i+1}(x_{i+1})$$

$$u_i = \bar{u}_i + \alpha_i + \beta_i(x_i - \bar{x}_i) \tag{4.3.4}$$

$$x_{i+1} = f_i(x_i, u_i)$$

for $i = 0, \dots, N-1$. Equation (4.3.4) may be rewritten as

$$V_i(\bar{x}+\delta x_i) = L_i(\bar{x}_i+\delta x_i, \bar{u}_i+\delta u_i) + V_{i+1}(\bar{x}_{i+1}+\delta x_{i+1}) \tag{4.3.5}$$

where

$$\delta x_{i+1} = f_i(\bar{x}_i+\delta x_i, \bar{u}_i+\delta u_i) - \bar{x}_{i+1} \tag{4.3.6}$$

If we now expand both sides of Equation (4.3.5) about the point $(\bar{x}_i, \bar{u}_i)$, we obtain

$$\begin{aligned} V_i(\bar{x}_i) + [V_x^{\,i}(\bar{x}_i)]^T \delta x_i + \tfrac{1}{2}\delta x_i^{\,T}[V_{xx}^{\,i}(\bar{x}_i)]\,\delta x_i &= L_i + V_{i+1}(\bar{x}_{i+1}) \\ &+ [H_x^{\,i}]^T \delta x_i + [H_u^{\,i}]\,\delta u_i + \tfrac{1}{2}\delta x_i^{\,T}[H_{xx}^i]\,\delta x_i + \delta u_i^{\,T}[H_{uu}^i]\,\delta x_i \\ &+ \tfrac{1}{2}\,\delta u_i^{\,T}[H_{uu}^i]\,\delta u_i + \tfrac{1}{2}\delta x_i^{\,T}[f_x^{\,i}]^T[V_{xx}^{i+1}(\bar{x}_{i+1})][f_x^{\,i}]\,\delta x_i \\ &+ \delta u_i^{\,T}[f_u^{\,i}]^T[V_{xx}^{i+1}(\bar{x}_{i+1})][f_x^{\,i}]\,\delta x_i \\ &+ \tfrac{1}{2}\delta u_i^{\,T}[f_u^{\,i}]^T[V_{xx}^{i+1}(\bar{x}_{i+1})][f_u^{\,i}]\,\delta u_i + e_i \end{aligned} \tag{4.3.7}$$

where the unspecified arguments are $\bar{x}_i$, $\bar{u}_i$, $V_x^{i+1}(\bar{x}_{i+1})$, $i = 0, \dots, N-1$. The error term e_i will be discussed later. For convenience, we define the following terms:

$$A_i = H_{xx}^i + [f_x^{\,i}]^T[V_{xx}^{i+1}(\bar{x}_{i+1})][f_x^{\,i}] \tag{4.3.8}$$

$$B_i = H_{ux}^i + [f_u^{\,i}]^T[V_{xx}^{i+1}(\bar{x}_{i+1})][f_x^{\,i}] \tag{4.3.9}$$

$$C_i = H_{uu}^i + [f_u^{\,i}]^T[V_{xx}^{i+1}(\bar{x}_{i+1})][f_u^{\,i}] \tag{4.3.10}$$

$$a_i = V_i(\bar{x}_i) - \bar{V}_i(\bar{x}_i) \tag{4.3.11}$$

for $i=0, \ldots, N-1$. $\bar{V}_i(\bar{x}_i)$ is the cost due to initial condition $\bar{x}_i$ and schedule $\{\bar{u}_i, \ldots, \bar{u}_{N-1}\}$, $V_i(\bar{x}_i)$ is the cost due to initial condition $\bar{x}_i$ and schedule $\{u_i, \ldots, u_{N-1}\}$, Hence $a_i(\bar{x}_i)$ is the change in cost. Using these definitions Equation (4.3.7) may be rewritten as

$$\begin{aligned} V_i(\bar{x}_i) + a_i + [V_x^i(\bar{x}_i)]^{\mathrm{T}} \delta x_i + \tfrac{1}{2}\delta x_i^{\mathrm{T}}[V_{xx}^i(\bar{x}_i)]\,\delta x_i \\ = L_i(\bar{x}_i, \bar{u}_i) + \bar{V}_{i+1}(\bar{x}_{i+1}) + a_{i+1} + [H_x^i]^T \delta x_i + [H_u^i]^T \delta u_i \qquad (4.3.12) \\ + \tfrac{1}{2}\delta x_i^T A_i \delta x_i + \delta u_i^T B_i \delta x_i + \tfrac{1}{2}\delta u_i^T C_i \delta u_i + e_i \end{aligned}$$

for $i=0, \ldots, N-1$. If we now replace δu_i by $\alpha_i + \beta_i \delta x_i$, and equate coefficients of like powers of δx_i, recalling that

$$\bar{V}_i(\bar{x}_i) = L_i(\bar{x}_i, \bar{u}_i) + \bar{V}_{i+1}(\bar{x}_{i+1}) \qquad (4.3.13)$$

we obtain

$$\begin{aligned} a_i &= a_{i+1} + [H_u^i]^T \alpha_i + \tfrac{1}{2}\alpha_i^T C_i \alpha_i + e_i' \\ V_x^i(\bar{x}_i) &= H_x^i + \beta_i^T H_u^i + [C_i \beta_i + B_i]^T \alpha_i + e_i'' \qquad (4.3.14) \\ V_{xx}^i(\bar{x}_i) &= A_i + \beta_i^T C_i \beta_i + \beta_i^T B_i + B_i^T \beta_i + e_i''' \end{aligned}$$

for $i=0, \ldots, N-1$, the unspecified arguments being $\bar{x}_i$, $\bar{u}_i$, $V_x^{i+1}(\bar{x}_{i+1})$. If we ignore the errors e_i', e_i'', e_i''', we obtain the following difference equations:

$$\begin{aligned} \hat{a}_i &= \hat{a}_{i+1} + [\hat{H}_u^i]^T \alpha_i + \tfrac{1}{2}\alpha_i^T \hat{C}\,\alpha_i \\ \hat{V}_x^i(\bar{x}_i) &= \hat{H}_x^i + \beta_i^T \hat{H}_u^i + [\hat{C}_i \beta_i + \hat{B}_i]^T \alpha_i \qquad (4.3.15) \\ \hat{V}_{xx}^i(\bar{x}_i) &= \hat{A}_i + \beta_i^T \hat{C}_i \beta_i + \beta_i^T \hat{B}_i + \hat{B}_i^T \beta_i \end{aligned}$$

where the unspecified arguments are $\bar{x}_i$, $\bar{u}_i$, and $\hat{V}_x^{i+1}(\bar{x}_{i+1})$, and the boundary conditions are

$$\hat{a}_N = 0$$

$$\hat{V}_x^N(\bar{x}_N) = F_x(\bar{x}_N), \qquad \hat{V}_{xx}^N(\bar{x}_N) = F_{xx}(\bar{x}_N) \qquad (4.3.16)$$

Because of the neglect of the error terms, $\hat{a}_i$, $\hat{V}_x^{\,i}(\bar{x}_i)$, and $\hat{V}_{xx}^{i}(\bar{x}_i)$, obtained by solving Equations (4.3.15) and (4.3.16), are estimates of the time parameters a_i, $V_x(\bar{x}_i)$, and $V_{xx}(\bar{x}_i)$ of the function $V_i(x_i)$:

$$a_i = V_i(\bar{x}_i) - \bar{V}_i(\bar{x}_i)$$

$$V_x^{\,i}(\bar{x}_i) = (\partial/\partial x)[V_i(\bar{x}_i)], \qquad V_{xx}^{i}(\bar{x}_i) = (\partial^2/\partial x^2)[V_i(\bar{x}_i)] \tag{4.3.17}$$

and $\hat{A}_i$, $\hat{B}_i$, and $\hat{C}_i$ are the corresponding values of A_i, B_i, and C_i. We must now determine the order of the error in these estimates. We assume that the third derivatives of $f_i(x_i, u_i)$, $L_i(x_i, u_i)$, $i=0, \ldots, N-1$, and $F(x_N)$, with respect to their arguments, exist and are bounded.

The error terms in Equation (4.3.14) may be found by evaluating the (previously neglected) third-order terms. The error terms, in the scalar case, have the following components:

e_i': $L_{uuu}^{i} \cdot \alpha_i^{3}$, $\quad V_x^{i+1} \cdot f_{uuu}^{i} \cdot \alpha_i^{3}$, $\quad V_{xx}^{i+1} \cdot f_u^{\,i} \cdot f_{uu}^{i} \cdot \alpha_i^{3}$

$V_{xxx}^{i+1} \cdot (f_u^{\,i})^3 \cdot \alpha_i^{3}$

e_i'': all the terms in e_i' with α_i^3 replaced by α_i^2

and: $L_{uux}^{i} \cdot \alpha_i^{2}$, $\quad V_x^{i+1} \cdot f_{uux}^{i} \cdot \alpha_i^{2}$, $\quad V_{xx}^{i+1} \cdot f_{uu}^{i} \cdot f_x^{\,i} \cdot \alpha_i^{2}$,

$V_{xx}^{i+1} \cdot f_u^{\,i} \cdot f_x^{\,i} \cdot \alpha_i^{2}$, $\quad V_{xxx}^{i+1} \cdot (f_u^{\,i})^2 \cdot f_x^{\,i} \cdot \alpha_i^{2}$

e_i''': all the terms in e_i'' with α_i^2 replaced by α_i

and: $L_{xxu}^{i} \cdot \alpha_i$, $\quad V_x^{i+1} \cdot f_{xxu}^{i} \cdot \alpha_i$, $\quad V_{xx}^{i+1} \cdot f_{xu} \cdot f_x \cdot \alpha_i$,

$V_{xxx}^{i+1} \cdot f_u^{\,i} \cdot (f_x^{\,i})^2 \cdot \alpha_i$

The unspecified argument of V_x^{i+1}, V_{xx}^{i+1}, V_{xxx}^{i+1} is $\bar{x}_{i+1}+\lambda(x_{i+1}-\bar{x}_{i+1})$. The unspecified arguments of the remaining terms are $\bar{x}_i$, $\bar{u}_i+\lambda(u_i-\bar{u}_i)$, where:

$$\begin{aligned} x_{i+1} &= f_i(\bar{x}_i, u_i) \\ u_i &= \bar{u}_i+\alpha_i \end{aligned} \tag{4.3.18}$$

and $0 \leqslant \lambda \leqslant 1$. If V_{xxx}^{i}, $i=1 \ldots N$, is bounded and α_i, $i=0 \ldots N-1$ is of order ε, then it follows from our assumptions that:

$$e_i' = O(\varepsilon^3), \qquad e_i'' = O(\varepsilon^2), \qquad e_i''' = O(\varepsilon) \tag{4.3.19}$$

Equations (4.3.15) are a set of coupled linear difference equations for the variables $\hat{a}_i$, $\hat{V}_x{}^i(\bar{x})$, $\hat{V}^i_{xx}(\bar{x}_i)$. Assume that the errors of the estimates $\hat{a}_{i+1}$, $\hat{V}_x^{i+1}(\bar{x}_{i+1})$, and $\hat{V}_{xx}^{i+1}(\bar{x}_{i+1})$ are of order ε^3, ε^2, and ε, respectively. Then, from Equations (4.3.8) to (4.3.10), the errors of the estimates $\hat{A}_i$, $\hat{B}_i$, and $\hat{C}_i$ are of order ε. Hence, from Equation (4.3.15), the error of the estimate $\hat{V}^i_{xx}(\bar{x}_i)$ is of order ε. Also the errors of $H_x{}^i$, $H_u{}^i$ and $[\hat{C}_i\beta_i+\beta_i]^T a_i$ are of order ε^2, so that the error of $\hat{V}_x{}^i(\bar{x}_i)$ is of order ε^2. Similarly, the error of $\hat{a}_i$ is of order ε^3. By induction [since the error of all the estimates $\hat{a}_N$, $\hat{V}_x{}^N(\bar{x}_N)$, $\hat{V}^N_{xx}(\bar{x}_N)$ is zero], it follows that

$$\begin{gathered} a_i - \hat{a}_i = O(\varepsilon^3) \\ V_x{}^i(\bar{x}) - \hat{V}_x{}^i(\bar{x}_i) = O(\varepsilon^2), \qquad V^i_{xx}(\bar{x}) - \hat{V}^i_{xx}(\bar{x}_i) = O(\varepsilon) \end{gathered} \tag{4.3.20}$$

for $i = 0, \ldots, N$.

For future convenience, we shall omit the circumflex, since it is implicitly understood that the resultant values of a_i, $V_x{}^i(\bar{x}_i)$, $V^i_{xx}(\bar{x}_i)$ are now estimates of the true parameters, with errors of order ε^3, ε^2, and ε, respectively. With this notation Equations (4.3.15) and (4.3.16) become,

$$\begin{aligned} a_i &= a_{i+1} + [H_u{}^i]^T \alpha_i + \tfrac{1}{2}\alpha_i{}^T C_i \alpha_i \\ V_x{}^i(\bar{x}_i) &= H_x{}^i + \beta_i{}^T H_u{}^i + [C_i\beta_i + B_i]^T \alpha_i \\ V^i_{xx}(\bar{x}_i) &= A_i + \beta_i{}^T C_i \beta_i + \beta_i{}^T B_i + B_i{}^T \beta_i \end{aligned} \tag{4.3.21}$$

$$\begin{gathered} a_N = 0 \\ V_x{}^N(\bar{x}_N) = 0, \qquad V^N_{xx}(\bar{x}_N) = 0 \end{gathered} \tag{4.3.22}$$

If $\varepsilon \to 0$, implying $\alpha_i \to 0$, $i = 0, \ldots, N-1$, the errors of all the estimates tend to zero; also $a_i \to 0$, $i = 0, \ldots, N-1$, and Equation (4.3.21) becomes identical to Equation (4.2.20) if $h_i(x_i)$ is replaced by the linear control law considered in this section, which leads to the replacement of $H_x{}'$ by $(H_x + \beta^T H_u)$ and the replacement of $(f_x{}^i)'$ by $(f_x{}^i + f_u{}^i \beta)$.

4.3.2. A Second-Order Optimization Algorithm for Unconstrained Discrete-Time Problems

We are now in a position to derive a second-order algorithm for determining at least a locally optimal trajectory. The algorithm generates an improved trajectory at each iteration. Suppose that the nominal sequences $\{\bar{u}_k\}$ and $\{\bar{x}_k\}$ are known, that a sequence of new control laws for the interval $[i+1, N-1]$ has been chosen, and that the resultant cost due to initial condition x_{i+1} and the new policy is $V_{i+1}(x_{i+1})$:

$$V_{i+1}(x_{i+1}) = \bar{V}_{i+1}(\bar{x}_{i+1}) + a_{i+1} + [V_x^{i+1}(\bar{x}_{i+1})]^T \delta x_{i+1} + \tfrac{1}{2}\delta x_{i+1}^T [V_{xx}^{i+1}(\bar{x}_{i+1})]^T \delta x_{i+1} + \cdots \tag{4.3.23}$$

where

$$\delta x_{i+1} = x_{i+1} - \bar{x}_{i+1} \tag{4.3.24}$$

We now formally apply the principle of optimality and choose δu_i to minimize $V_i(x_i)$. Later we show how the magnitude of δu_i may be restricted. Thus,

$$V_i(x_i) = \min_{\delta u_i} [L_i(\bar{x}_i + \delta x_i, \bar{u}_i + \delta u_i) + V_{i+1}(x_{i+1})] \tag{4.3.25}$$

where

$$x_{i+1} = f_i(\bar{x}_i + \delta x_i, \bar{u}_i + \delta u_i) \tag{4.3.26}$$

Expanding each side of Equation (4.3.25), up to second-order terms, about the point $(\bar{x}_i, \bar{u}_i)$, we obtain†

$$\begin{aligned}\bar{V}_i(\bar{x}_i) + a_i + [V_x^i(\bar{x}_i)]^T \delta x_i + \tfrac{1}{2}\delta x_i^T [V_{xx}^i(\bar{x}_i)] \delta x_i \\ = \min_{\delta u_i} [L_i(\bar{x}_i, \bar{u}_i) + \bar{V}_{i+1}(\bar{x}_{i+1}) + a_{i+1} + [H_x^i]^T \delta x_i + [H_u^i]^T \delta u_i \\ + \tfrac{1}{2}\delta x_i^T A_i \delta x_i + \delta u_i^T B_i \delta x_i + \tfrac{1}{2}\delta u_i^T C_i \delta u_i + e_i]\end{aligned} \tag{4.3.27}$$

If we ignore the error e_i and the restriction on the magnitude of δu_i, the optimal δu_i may be obtained by differentiating the right side of Equation (4.3.27) with respect to δu_i:

$$C_i \delta u_i + H_u^i + B_i \delta x_i = 0 \tag{4.3.28}$$

† See Equation (4.3.7).

If C_i is positive-definite, the unique minimum† of the right-hand side of Equation (4.3.27) is given by

$$\delta u_i = \alpha_i + \beta_i \delta x_i \tag{4.3.29}$$

where

$$\alpha_i = -[C_i]^{-1} H_u^{\,i} \tag{4.3.30}$$

$$\beta_i = -[C_i]^{-1} B_i \tag{4.3.31}$$

However the various estimates we require are only accurate if α_i, $i=0, \ldots, N-1$, is sufficiently small. One way of achieving this is to replace Equation (4.3.30) by

$$\alpha_i = -\varepsilon[C_i]^{-1} H_u^{\,i}, \qquad 0<\varepsilon \leqslant 1 \tag{4.3.32}$$

ensuring (if C_i is positive-definite) that α_i is of order ε. Substituting these values of α_i and β_i into Equation (4.3.21) yields the following difference equations for a_i, $V_x^{\,i}(\bar{x}_i)$, $V_{xx}^i(\bar{x}_i)$:

$$\begin{aligned} a_i &= a_{i+1} - \varepsilon(1-\varepsilon/2)[H_u^{\,i}]^T[C_i]^{-1}[H_u^{\,i}] \\ V_x^{\,i}(\bar{x}_i) &= H_x^{\,i} + \beta_i^T H_u^{\,i} \\ V_{xx}^i(\bar{x}_i) &= A_i - \beta_i^T C_i \beta_i \end{aligned} \tag{4.3.33}$$

for $i=0, \ldots, N-1$. The boundary conditions are given by Equation (4.3.22). Recalling the discussion of Section 4.3.1, it can be seen that a_i, $V_x^{\,i}(\bar{x}_i)$, and $V_{xx}^i(\bar{x}_i)$, obtained as solutions of Equation (4.3.33) are estimates of the true parameters of $V_i(x_i)$, the cost due to initial condition x_i and the new policy defined by Equations (4.3.29)-(4.3.31). Provided C_i is invertible (not necessarily positive-definite), the estimated change in cost of the new policy is

$$a_0 = -\varepsilon(1-\varepsilon/2) \sum_{i=0}^{N-1} [H_u^{\,i}]^T[C_i]^{-1}[H_u^{\,i}] \tag{4.3.34}$$

† Ignoring e_i.

Hence,

$$a_o = O(\varepsilon) \tag{4.3.35}$$

except when $\{\bar{u}_i\}$ is optimal. If, in addition, C_i is positive-definite, $(i=0, \ldots, N-1)$, then a_o, the estimated change in cost, is negative. Let ΔV_o denote the true change in cost. Then

$$|\Delta V_o - a_o| = O(\varepsilon^3) \tag{4.3.36}$$

From Equations (4.3.35) and (4.3.36) there exists an ε, sufficiently small, such that $a_o < 0$ gurantees $\Delta V_o < 0$. For this value of ε, the new policy produces a guaranteed reduction in cost.

By formally applying the principle of optimality, we have obtained a new policy defined by Equations (4.3.29)-(4.3.31). This policy is not optimal, because of the neglect of e_i in the minimization, the constraint on α_i, and the errors in the estimates C_i, $H_u{}^i$ and B_i. Nevertheless, adoption of this policy (if C_i is positive-definite, $i=0, \ldots, N-1$, and the third derivatives of f_i, L_i, $i=0, \ldots, N-1$, and F exist and are bounded) results in a guaranteed reduction in cost for any nonoptimal nominal schedule $\{\bar{u}_i\}$. An estimate, with error $o(\varepsilon^3)$, of the change in cost, is given by Equation (4.3.34). The change is of order ε. Note that optimality implies that $H_u{}^i = 0$, $i=0, \ldots, N-1$.

4.3.3. The Algorithm

The algorithm may now be stated:

1. For given initial condition $\bar{x}_o$ and schedule $\{\bar{u}_i\}$, calculate $\{\bar{x}_i\}$. Set $\varepsilon = 1$.

2. Calculate $\{a_i\}$, $\{V_x{}^i(\bar{x}_i)\}$, and $\{V_{xx}^i(\bar{x}_i)\}$ using Equation (4.3.33) [using $\varepsilon = 1$ for $\{a_i\}$; a_i, $i=0, \ldots, N$, for any other value of ε can be obtained by multiplying the value for $\varepsilon = 1$ by $\varepsilon(1-\varepsilon/2)$] and the boundary conditions of Equation (4.3.22). Store $\alpha_i = -[C_i]^{-1} H_u{}^i$, $\beta_i = -[C_i]^{-1} B_i i=0, \ldots, N-1$.

3. For given ε, calculate the new control and state sequences using

$$x_{i+1} = f_i(\bar{x}_i, u_i), \qquad x_o = \bar{x}_o$$

$$u_i = \bar{u}_i + \varepsilon\alpha_i + \beta_i(x_i - \bar{x}_i)$$

4. Calculate the actual change in cost:

$$\Delta V_0 = \sum_{i=1}^{N-1} [L_i(x_i, u_i) - L_i(\bar{x}_i, \bar{u}_i)] + F(x_N) - F(\bar{x}_N)$$

If ΔV_0 is positive (i.e., the new policy has increased the cost), or if ΔV_0 is negative, but

$$\frac{|\Delta V_0|}{|\varepsilon(1-\varepsilon/2)a_0|} < c, \qquad 0 < c \leqslant 1$$

[$\varepsilon(1-\varepsilon/2)a_0$ is the estimated value of ΔV_0], set $\varepsilon = \varepsilon/2$ and repeat Step 3†. If $\Delta V_0 < 0$ and

$$\frac{|\Delta V_0|}{|\varepsilon(1-\varepsilon/2)a_0|} > c$$

the new sequences $\{x_i\}$ and $\{u_i\}$ are accepted and stored.

5. Set $\bar{u}_i = u_i$, $i = 0, \ldots, N-1$; $\bar{x}_i = x_i$, $i = 0, \ldots, N$, and repeat Step 1.
6. Computation is halted when

$$|a_0(\bar{x}_0)| \leqslant \eta$$

where η is a small, positive quantity determined from numerical stability considerations.

Several adaptations of the above basic algorithm may be employed. Instead of Step 4, ε may be chosen to minimize the new cost $V_0(\bar{x}_0)$. If C_i is not positive-definite, $i = 0, \ldots, N-1$, a_0 is not necessarily negative, so that a reduction in cost for ε sufficiently small is not guaranted. Various devices may be used to avoid this difficulty. If at time i, $[H_u^i]^T [C_i]^{-1} [H_u^i] < 0$, set

$$\delta u_i = -\varepsilon_i H_u^i + \beta_i \delta x_i, \qquad \varepsilon_i > 0$$

and change the difference equation for a_i at time i only, to

$$a_i = a_{i+1} - \varepsilon_i [H_u^i]^T [H_u^i]$$

This guarantees a reduction in cost for ε_i sufficiently small. However ε_i has to be chosen.

† $c = 0.5$ has been found to be satisfactory.

4.3.4. Properties of the Algorithm

The most important property of the algorithm is that it provides one step convergence for LQP problems (linear systems, quadratic performance index). This can be proved quite simply by noting that the algorithm reduces to the conventional dynamic programming solution of LQP problems except that the origin of the state and control spaces is taken to be $\bar{x}_i$, $i=0, \ldots, N$, and $\bar{u}_i$, $i=0, \ldots, N-1$. In particular, the difference equation for $V_{xx}^i(\bar{x}_i)$ becomes the matrix Riccati difference equation. An algorithm, based on the calculus of variations, would not reduce to the conventional dynamic programming solution when applied to the LQP problem. Coupled with this is the fact that there is one set of vector difference equations less in the differential dynamic programming algorithm than would be required by an algorithm based on the calculus of variations. This has been pointed out in Chapter 2.

We have not yet discussed the convergence of the algorithm. Using a theorem by Polak [3], it is possible to prove that the algorithm converges to a schedule $\{u_k\}$ satisfying the necessary condition of optimality $H_u^{\,i} = 0$, $i = 0, \ldots, N-1$.

4.3.5. An Example

Consider the following discrete-time version of a continuous-time problem considered by Fuller [4]. The system equations are

$$x_{i+1} = x_i + Ts(u_i)$$

$$y_{i+1} = y_i + Tx_i + (T^2/2)\, s(u_i)$$

where $s(\cdot)$ is the saturating function defined by

$$s(u) = u, \qquad |u| \leqslant D$$

$$s(u) = 1 - (1-D)\exp\left[\frac{-u+D}{1-D}\right], \qquad u > D$$

$$s(u) = -1 + (1-D)\exp\left[\frac{u+D}{1-D}\right], \qquad u < -D$$

As $T \to 0$ and $D \to 1$, the above equations tend to

$$\dot{x} = \operatorname{sat}(u)$$

$$\dot{y} = x$$

where

$$\text{sat}(u) = u, \qquad |u| \leqslant 1$$

$$\text{sat}(u) = 1, \qquad u > 1$$

$$\text{sat}(u) = -1, \quad u < -1$$

The cost function is

$$V_o = \sum_{i=0}^{N-1} (T/2)(Ru_i^2 + y_i^2) + (T/2)\, y_N^2$$

As $T \to 0$ and $R \to 0$, V_o tends to

$$V_o = \int_0^{t_f} \tfrac{1}{2} y^2 \, dt$$

where

$$t_f = NT$$

For the numerical experiment, N was chosen to be 1000, T to be 0.01 sec, making $t_f = 10$ sec. The number of iterations required to obtain the optimal increased as R decreased as follows:

$D = 0.5, \quad R = 0.1:$ 5 iterations

$D = 0.5, \quad R = 0.01:$ 6 iterations

$D = 0.5, \quad R = 0.001:$ 10 iterations

$D = 0.5, \quad R = 0.0001:$ 20 iterations

The resultant optimal trajectories are shown in Figure 4.1. As R decreases, the "corner" in the trajectory (denoting change of sign in u_i) approaches more closely to the switching curve, defined by $y = -0.4446x|x|$, obtained by Fuller [4] for the continuous-time problem. Obviously the continuous-time problem is better treated by the method given in Chapter 3.

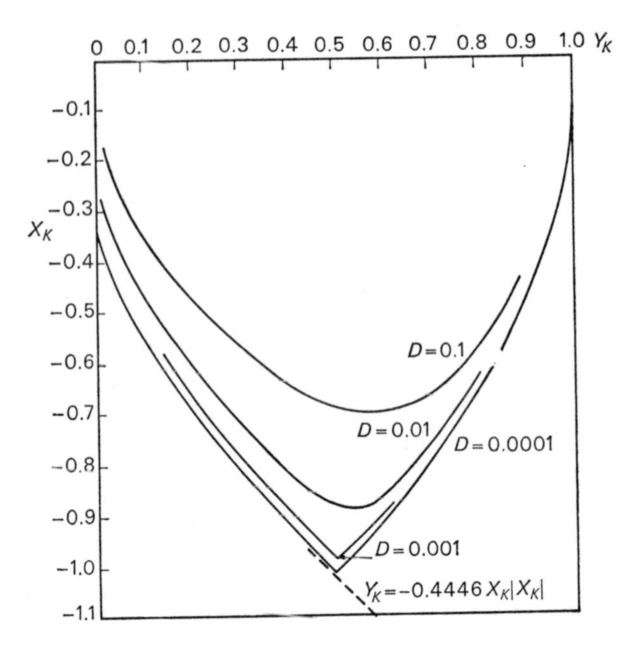

Figure 4.1

4.3.6. First-Order Algorithm

The first-order algorithm can be obtained very simply using the procedure of Section 4.3.2, by neglecting terms of second-order and above. Thus, the local control law:

$$\delta u_i = -\varepsilon_i H_u^{\,i} \tag{4.3.37}$$

yields the following iterative equations:

$$\begin{aligned} a_i &= a_{i+1} - \varepsilon_i [H_u^{\,i}]^T [H_{u_i}] \\ V_x^{\,i}(\bar{x}_i) &= H_x^{\,i} \end{aligned} \tag{4.3.38}$$

with boundary conditions given, as before, by Equations (4.3.22) $\varepsilon_i [H_u^{\,i}]^T [H_u^{\,i}]$ is the estimated reduction in cost at time i, and is clearly positive if $H_u^{\,i} \neq 0$. Also a_o, where

$$a_o = -\sum_{i=0}^{N-1} \varepsilon_i [H_u^{\,i}]^T [H_{u_i}] \tag{4.3.39}$$

is the total estimated change in cost using the revised control policy defined by Equation (4.3.37), $i=0, \ldots, N-1$. It is often convenient to set $\varepsilon_0 = \varepsilon_2 = \ldots = \varepsilon_{N-1} = \varepsilon$. The solution of Equations (4.3.38) yields estimates a_i and $V_x^{\ i}$ with errors of $0(\varepsilon^2)$ and $0(\varepsilon)$ respectively. A suitable algorithm is

1. For given initial condition $\bar{x}_0$ and sequence $\{\bar{u}_i\}$ calculate $\{\bar{x}_i\}$.

2. In reverse time calculate $\{a_i\}$ and $\{V_x^{\ i}(\bar{x}_i)\}$ using Equations (4.3.38), and the boundary conditions given by Equations (4.3.22).

3. For given ε, calculate new control and state sequences using

$$x_{i+1} = f_i(x_i, u_i), \qquad x_0 = \bar{x}_0$$

$$u_i = \bar{u}_i - \varepsilon H_u^{\ i}$$

4. Calculate the actual change in cost ΔV^0 and compare with the estimated reduction $|a_0|$. If ΔV^0 is negative and $|\Delta V^0|/|a_0| > c$, ($c=0.5$ usually), accept the new trajectory. Otherwise reduce ε and repeat Step 3.

5. Set $\bar{x}_i = x_i$ ($i=1, \ldots, N$), $u_i = \bar{u}_i$, reset ε, and repeat Step 2.

6. Stop when $|a_0| < \eta$.

4.4. DIFFERENTIAL DYNAMIC PROGRAMMING—GLOBAL VARIATIONS IN CONTROL

4.4.1. The Scalar Case

We have managed in the previous sections of this chapter to avoid notational difficulties [such as the notation required for the second-derivative of a vector function $f(x)$ with respect to the vector x] by the introduction of the Hamiltonian function H. These difficulties cannot be avoided when global variations in control are permitted. To avoid these notational difficulties hindering the exposition, it is preferable to consider initially the case where x and u are scalars.

The starting point is again the relation:

$$V_i(\bar{x}_i+\delta x_i) = L_i(\bar{x}_i+\delta x_i, u_i) + V_{i+1}(\bar{x}_{i+1}+\delta x_{i+1}) \qquad (4.4.1)$$

where

$$\delta x_{i+1} = x_{i+1} - \bar{x}_{i+1} = f_i(\bar{x}_i+\delta x_i, u_i) - f_i(\bar{x}_i, \bar{u}) \qquad (4.4.2)$$

As before, it is assumed that $V_{i+1}(x_{i+1})$ is known in the neighborhood of $\bar{x}_{i+1}$. In Section 4.2 we restricted u_i to the neighborhood of $\bar{u}_i$, i.e., we set

$$u_i = \bar{u}_i + a_i + \beta_i \delta x_i$$

where α_i was suitably restricted. This enabled us to approximate $L_i(\bar{x}_i, u_i)$ and $f_i(\bar{x}_i, u_i)$ by Taylor series expansions about the point $(\bar{x}_i, \bar{u}_i)$. However, in many problems large variations in u_i cause small variations in x_{i+1}. Thus, the Euler discretization of

$$\dot{x} = \tilde{f}(x, u, t)$$

is

$$x_{i+1} = x_i + \tilde{f}(x_i, u_i, iT)T$$

so that

$$f_i(\bar{x}_i, u_i) = \bar{x}_i + \tilde{f}(\bar{x}_i, u_i, iT)T$$

Now $V_{i+1}(x_{i+1})$ is known (to a specified degree of accuracy) in a neighborhood of $\bar{x}_{i+1}$. To reach the boundary of the neighborhood, from a given state $\bar{x}_i$ might, if T is small, require a large δu_i. In Section 4.2 $f_i(\bar{x}_i, u_i)$ was estimated by a second-order Taylor series expansion of $f_i(\bar{x}_i, u_i)$ about the point $(\bar{x}_i, \bar{u}_i)$. The restriction on $\|\delta u_i\|$, in order to ensure the accuracy of this estimate, might confine $f_i(\bar{x}_i, u_i)$ to a much smaller neighborhood of $\bar{x}_{i+1}$ than the one for which $V_{i+1}(x_{i+1})$ is known. Since this knowledge of $V_{i+1}(x_{i+1})$, i.e., the parameters a_{i+1}, $V_x^{i+1}(\bar{x}_{i+1})$, and $V_{xx}^{i+1}(\bar{x}_{i+1})$, is computationally expensive to obtain, it is necessary to obtain an estimate of x_{i+1} that permits relatively large variations in u_i. Let us therefore consider the effect of a global change in u_i to u_i^*, and let δu_i now be the change in u_i from u_i^*:

$$u_i = u_i^* + \delta u_i \tag{4.4.3}$$

Then

$$\begin{aligned}
\delta x_{i+1} &= x_{i+1} - \bar{x}_{i+1} \\
&= f_i(\bar{x}_i + \delta x_i, u_i^* + \delta u_i - f_i(\bar{x}_i, \bar{u}_i) \\
&= f_i(\bar{x}_i, u_i^*) - f_i(\bar{x}, \bar{u}_i) + f_x^i \delta x_i + f_u^i \delta u_i + \tfrac{1}{2} f_{xx}^i \delta x_i^2 \\
&\quad + f_{ux}^i \delta u_i \delta x_i + \tfrac{1}{2} f_{uu}^i \delta u_i^2 + \cdots
\end{aligned} \tag{4.4.4}$$

where the unspecified arguments are x_i, u_i^* (*not* $\bar{x}_i$, $\bar{u}_i$ as in Section 4.2). Let us define Δf_i as

$$\Delta f_i = f_i(\bar{x}_i, u_i^*) - f_i(\bar{x}_i, \bar{u}_i) \tag{4.4.5}$$

so that

$$\delta x_{i+1} = \Delta f_i + f_x{}^i \delta x_i + f_u{}^i \delta u_i + \tfrac{1}{2} f^i_{xx} \delta x_i{}^2 + f^i_{ux} \delta u_i \delta x_i + \tfrac{1}{2} f^i_{xx} \delta x_i{}^2 + \cdots \tag{4.4.6}$$

Similarly,

$$L_i(\bar{x}_i + \delta x_i, u_i^* + \delta u_i) = L_i(\bar{x}_i, \bar{u}_i) + \Delta L_i + L_x{}^i \delta x_i + L_u{}^i \delta u_i$$
$$+ \tfrac{1}{2} L^i_{xx} \delta x_i{}^2 + L^i_{ux} \delta u_i \delta x_i + \tfrac{1}{2} L^i_{uu} \delta u_i{}^2 + \cdots \tag{4.4.7}$$

where

$$\Delta L_i = L_i(\bar{x}_i, u_i^*) - L_i(\bar{x}_i, \bar{u}_i) \tag{4.4.8}$$

Substituting Equations (4.4.7) into Equation (4.4.1), using the appropriate second-order expansion of V_i and V_{i+1} yields:

$$\bar{V}_i(\bar{x}_i) + a_i + V_x{}^i(\bar{x}_i) \delta x_i + \tfrac{1}{2} V^i_{xx}(\bar{x}_i) \delta x_i{}^2 + \cdots$$
$$= L_i(\bar{x}_i, \bar{u}_i) + \bar{V}_{i+1}(\bar{x}_{i+1}) + \Delta L_i + L_x{}^i \delta x_i + L_u{}^i \delta u_i + \tfrac{1}{2} L^i_{xx} \delta x_i{}^2$$
$$+ L^i_{ux} \delta u_i \delta x_i + \tfrac{1}{2} L^i_{uu} \delta u_i{}^2 + a_{i+1} + V_x^{i+1}(\bar{x}_{i+1}) \delta x_{i+1} \tag{4.4.9}$$
$$+ \tfrac{1}{2} V_{xx}^{i+1}(\bar{x}_{i+1}) \delta x_{i+1}^2 + \cdots$$

where δx_{i+1} is given by Equation (4.4.6), and the unspecified arguments are $\bar{x}_i$ and u_i^*. Substituting Equation (4.4.7) into Equation (4.4.9) and making use of the relation:

$$\bar{V}_i(\bar{x}_i) = L_i(\bar{x}_i, \bar{u}_i) + \bar{V}_{i+1}(\bar{x}_{i+1})$$

yields

$$a_i + V_x{}^i(\bar{x}_i) \delta x_i + \tfrac{1}{2} V^i_{xx}(\bar{x}_i) \delta x_i{}^2$$
$$= a_{i+1} + [\Delta L_i + V_x^{i+1} \Delta f_i] + \tfrac{1}{2} V_{xx}^{i+1} (\Delta f_i)^2$$

$$+ [L_x^{\,i} + V_x^{i+1} f_x^{\,i} + V_{xx}^{i+1}\, \Delta f_i\, f_x^{\,i}]\, \delta x_i + \tfrac{1}{2}[L_{xx}^i + V_x^{i+1} f_{xx}^i$$

$$+ \tfrac{1}{2} V_{xx}^{i+1} (f_x^{\,i})^2]\, \delta x_i^{\,2} + \cdots \tag{4.4.10}$$

$$= a_{i+1} + \Delta H_i + \tfrac{1}{2} V_{xx}^{i+1} (\Delta f_i)^2 + [H_x^{\,i} + V_{xx}^{i+1}\, \Delta f_i\, f_x^{\,i}]\, \delta x_i$$

$$+ [H_u^{\,i} + V_{xx}^{i+1}\, \Delta f_i\, f_u^{\,i}]\, \delta u_i + \tfrac{1}{2}[H_{xx}^i + V_{xx}^{i+1} (f_x^{\,i})^2 + V_{xx}^{i+1}\, \Delta f_i\, f_{xx}^i]\, \delta x_i^{\,2}$$

$$+ [H_{ux}^i + V_{xx}^{i+1} f_u^{\,i} f_x^{\,i} + V_{xx}^{i+1}\, \Delta f_i\, f_{ux}^i]\, \delta u_i\, \delta x_i$$

$$+ \tfrac{1}{2}[H_{uu}^i + V_{xx}^{i+1} (f_u^{\,i})^2 + V_{xx}^{i+1}\, \Delta f_i\, f_{uu}^i]\, \delta u_i^{\,2} + e_i$$

where the unspecified argument of $V_x^{\,+1}$ and V_{xx}^{i+1} is $\bar{x}_{i+1}$, and the unspecified arguments of the remaining terms are $\bar{x}_i$, u_i^*, $V_x^{i+1}(\bar{x}_{i+1})$. Also:

$$\Delta H_i = H_i(\bar{x}_i, u_i^*, V_x^{i+1}(\bar{x}_{i+1})) - H_i(\bar{x}_i, \bar{u}_i, V_x^{i+1}(\bar{x}_{i+1})) \tag{4.4.11}$$

e_i denotes the error term. For convenience, let us define A_i, B_i, and C_i as follows [note that these definitions differ from those used in Section 4.2 since $(\bar{x}_i, u_i^*)$ replaces $(\bar{x}_i, \bar{u}_i)$, and extra terms involving Δf_i are introduced]:

$$A_i = H_{xx}^i + V_{xx}^{i+1}[(f_x^{\,i})^2 + \Delta f_i\, f_{xx}^i] \tag{4.4.12}$$

$$B_i = H_{ux}^i + V_{xx}^{i+1}[f_u^{\,i} f_x^{\,i} + \Delta f_i\, f_{ux}^i] \tag{4.4.13}$$

$$C_i = H_{uu}^i + V_{xx}^{i+1}[(f_u^{\,i})^2 + \Delta f_i\, f_{uu}^i] \tag{4.4.14}$$

where the unspecified arguments are x_i, u_i^*, $V_x^{i+1}(\bar{x}_{i+1})$. Equation (4.4.10) becomes

$$\begin{aligned} a_i + V_x^{\,i}(\bar{x}_i)\, \delta x_i + \tfrac{1}{2} V_{xx}^i(\bar{x}_i)\, \delta x_i^{\,2} = {} & a_{i+1} + \Delta H_i + \tfrac{1}{2} V_{xx}^{i+1}(\bar{x}_{i+1})(\Delta f_i)^2 \\ & + [H_x^{\,i} + V_{xx}^{i+1}(\bar{x}_{i+1})\, \Delta f_i\, f_x^{\,i}]\, \delta x_i \\ & + [H_u^{\,i} + V_{xx}^{i+1}(\bar{x}_{i+1})\, \Delta f_i\, f_u^{\,i}]\, \delta u_i \\ & + \tfrac{1}{2} A_i\, \delta x_i^{\,2} + B_i\, \delta u_i\, \delta x_i + \tfrac{1}{2} C_i\, \delta u_i^{\,2} \\ & + e_i \end{aligned} \tag{4.4.15}$$

We can now consider the various cases considered in Section 4.2.

Linear nonoptimal control policy: consider the policy defined by :

$$u_i = u_i^* + \delta u_i, \qquad \delta u_i = \beta_i \delta x_i \tag{4.4.16}$$

for $i = 0, \ldots, N-1$. Substituting Equation (4.4.16) into Equation (4.4.15), equating coefficients of like powers of δx_i, and neglecting the error term, yields the following iterative equations of a_i, $V_x^i(\bar{x}_i)$ and $V_{xx}^i(\bar{x}_i)$:

$$\begin{aligned} a_i &= a_{i+1} + \Delta H_i + \tfrac{1}{2} V_{xx}^{i+1}(\bar{x}_{i+1})(\Delta f_i)^2 \\ V_x^i(\bar{x}_i) &= H_x^i + \beta_i H_u^i + [f_x^i + f_u^i \beta_i] V_{xx}^{i+1}(\bar{x}_{i+1}) \Delta f_i \\ V_{xx}^i(\bar{x}_i) &= A_i + 2 B_i \beta_i + C_i \beta_i^2 \end{aligned} \tag{4.4.17}$$

The unspecified arguments are $\bar{x}_i$, u_i^*, and $V_x^{i+1}(\bar{x}_{i+1})$. The boundary conditions remain the same as those given in Section 4.2.

Linear optimal control policy: if we wish to minimize the right side of Equation (4.4.15), neglecting e_i for the moment, we encounter a certain ambiguity, since $u_i = u_i^* + \delta u_i$, and both u_i^* and δu_i have to be chosen. We can, in fact, choose u_i^* arbitrarily, and then choose δu_i to minimize the right-hand side. However, it is sensible to choose u_i^* to minimize the right-hand side with $\delta u_i = \delta x_i = 0$. Then u_i^* is the (estimated) optimal control action for the state $\bar{x}_i$. We then choose δu_i to minimize the right-hand side, and this yields δu_i as a function of δx_i, and, hence, the (estimated) optimal control action for x_i is the neighborhood of $\bar{x}_i$. Thus, in the sequel, u_i^* is defined as follows:

$$u_i^* = u_i \quad \text{such that} \quad H_i(\bar{x}_i, u_i, V_x^{i+1}(\bar{x}_{1+1})) - H_i(\bar{x}_i, \bar{u}_i, V_x^{i+1}(\bar{x}_{i+1})) + \tfrac{1}{2} V_{xx}^{i+1}(\bar{x}_{i+1})[f_i(\bar{x}_i, u_i) - f_i(\bar{x}_i, \bar{u}_i)]^2 \quad \text{is minimum} \tag{4.4.18}$$

or

$$u_i^* = u_i \quad \text{such that} \quad H_i(\bar{x}_i, u_i, V_x^{i+1}(\bar{x}_{i+1})) + \tfrac{1}{2} V_{xx}^{i+1}(\bar{x}_{i+1})[f_i(\bar{x}_i, u_i) - f_i(\bar{x}_i, \bar{u}_i)]^2 \quad \text{is minimum} \tag{4.4.19}$$

At $u_i = u_i^*$ the following relation is satisfied:

$$H_u^i + V_{xx}^{i+1}(\bar{x}_{i+1}) \Delta f_i f_u^i = 0 \tag{4.4.20}$$

Note that u_i^* does not necessarily minimize H_i. Substituting u_i^*, defined by Equation (4.4.18) or (4.4.19) into Equation (4.4.15), and differentiating with respect to δu_i yields the following condition for the optimal δu_i (neglecting e_i):

$$[H_u^i + V_{xx}^{i+1}(\bar{x}_{i+1})\Delta f_i f_u^i] + B_i \delta x_i + C_i \delta u_i = 0 \tag{4.4.21}$$

where the unspecified arguments are $\bar{x}_i$, u_i^*, and $V_x^{i+1}(\bar{x}_{i+1})$. The first (bracketed) term in Equation (4.4.21) is zero. Hence, if C_i is positive, the (unique) δu_i that minimizes the right-hand side of Equation (4.4.15) is

$$\delta u_i = \beta_i \delta x_i \tag{4.4.22}$$

where

$$\beta_i = -C_i^{-1} B_i \tag{4.4.23}$$

Substituting Equations (4.4.22) and (4.4.23) into the difference equations (4.4.17), and recalling the definitions of A_i, B_i, and C_i, yields

$$\begin{aligned} a_i &= a_{i+1} + \Delta H_i + \tfrac{1}{2} V_{xx}^{i+1}(\bar{x}_{i+1})(\Delta f_i)^2 \\ V_x^i(\bar{x}_i) &= H_x^i + f_x^i V_{xx}^{i+1}(\bar{x}_{i+1})\Delta f_i \\ V_{xx}^i(\bar{x}_i) &= A_i - C_i \beta_i^2 \end{aligned} \tag{4.4.24}$$

[The unspecified arguments are $\bar{x}_i$, u_i^*, $V_x^{i+1}(\bar{x}_{i+1})$.]

The estimation error: both Equations (4.4.17) and (4.4.18) were obtained by neglecting certain error terms. The variables a_i, $V_x^i(\bar{x}_i)$, and $V_{xx}^i(\bar{x}_i)$ must, therefore, be regarded as estimates of the true parameters defining $V_i(x_i)$ in the neighborhood of $\bar{x}_i$. We make the same assumptions as before, namely, that the third derivatives of $f_i(x_i, u_i)$, $L_i(x_i, u_i)$, $i = 0, \ldots, N-1$, and $F(x_N)$, with respect to their arguments, exist and are bounded. The errors e_i', e_i'', e_i''' in the difference equations for a_i, V_x^i, V_{xx}^i are composed of the following terms†:

$$e_i' : V_{xxx}^{i+1}(\Delta f_i)^3, \qquad e_i'' : V_{xxx}^{i+1}(\Delta f_i)^2 (f_x^i)'$$

$$e_i''' : V_{xxx}^{i+1}\Delta f_i[(f_x^i)']^2, \qquad (f_{xx}^i + \beta_i f_{ux}^i) V_{xx}^{i+1}\Delta f_i$$

† $(f_x^i)' = f_x^i + f_u^i \beta_i$.

etc. The error terms are fewer in number than in Section 4.3.1, since the point of expansion is $(\bar{x}_i, u_i^*)$. If $|u_i^* - \bar{u}_i|$ is of order ε, Δf_i is of order ε and hence, as in Section 4.3.1,

$$e_i' = 0(\varepsilon^3), \qquad e_i'' = 0(\varepsilon^2), \qquad e_i''' = 0(\varepsilon)$$

Equation (4.4.17) and (4.4.24) are linear difference equations in the variables a_i, $V_x^{\,i}$, and V_{xx}^i. Hence, as before, the solution of Equations (4.4.17) and (4.4.24) yields estimates a_i, $V_x^{\,i}$, and V_{xx}^i of the true parameters, the errors in these estimates being of order, respectively, ε^3, ε^2, and ε, *provided that* $|u_i^* - \bar{u}_i|$ is of order ε for $i = 0, \ldots, N-1$. Under this condition, since ΔH_i is then also of order ε, a_0 (the estimated change in cost) is also of order ε for $\{\bar{u}_i\}$ nonoptimal. Since u_i^* minimizes Equation (4.4.18) with respect to u_i, a_0 is always negative. The true change in cost differs from a_0 by $0(\varepsilon^3)$. If a_0 is negative (see below for sufficient conditions to ensure this) there exists an ε, sufficiently small, such that the true change in cost is negative.

Step-size adjustment procedure. A reduction of cost with each iteration of the algorithm can only be guaranteed if $|u_i^* - u_i|$ is of order ε, $i = 0, \ldots, N-1$, and ε is sufficiently small. In the algorithm of Section 4.2, ε could be specified and $u_i^* - u_i$ (α_i) chosen to be of order ε. In the algorithm discussed here, u_i^* is chosen by optimizing Equation (4.4.18) and only if u_i is sufficiently close to the optimal control u_i° will $|u_i - u_i^*|$ be sufficiently small. If, therefore, $\{u_i\}$ is not sufficiently close to $\{u_i^{\circ}\}$, another method of effectively choosing the step size has to be employed.

This can be done essentially by the technique employed in Chapter 2, with one extra feature. Essentially the new policy is applied over a portion of the interval $[0, N-1]$, and this portion is reduced until an improved cost is obtained. If an improvement is still not obtained when the portion has been reduced to one sampling period, the control action for this period is changed until an improvement is obtained.

For the given sequences $\{\bar{x}_i\}$ and $\{\bar{u}_i\}$, there will exist a time $N_{\text{eff}} \leqslant N$ such that $a_{N_{\text{eff}}} < 0$, but $a_i = 0$, $i > N_{\text{eff}}$. This means that the trajectory over the interval $[N_{\text{eff}} + 1, N]$ is optimal, hence improvement of the cost must be effected by changes over the interval $[0, N_{\text{eff}}]$. The estimated optimal policy is defined by

$$u_i = u_i^* + \beta_i \, \delta x_i, \qquad i = 0, \ldots, N_{\text{eff}}$$

where u_i^* is given by Equation (4.4.18) and β_i by Equation (4.4.23). Also,

$$u_i^* = 0, \qquad i = (N_{\text{eff}} + 1), \ldots, N-1$$

The new policy is applied, not over the whole interval $[0, N-1]$, but over the interval $[N_1, N_{\text{eff}}]$, and this interval is reduced (by successively halving the interval $[N_1, N_{\text{eff}}]$) until an acceptability criterion is satisfied. The criterion could be, as before, that the actual change ΔV_0 in cost is negative and $|\Delta V_0|$ is at least $\frac{1}{2}|a_0|$, a_0 being the expected change in cost. If this criterion is still not satisfied for $N_1 = N_{\text{eff}}$, then u^*_{N-1} is chosen, not to minimize Equation (4.4.18), but according to an algorithm that allows us to ensure that $|u^*_{N_1} - \bar{u}_{N_1}|$ is of order ε, ε prespecified, and that the appropriately modified value of a_{N_1} is negative. For example, if $N_1 = N_{\text{eff}}$,

$$u^*_{N_1} = u_{N_1} - \varepsilon H_u^{N_1}, \qquad \varepsilon > 0$$

whence

$$\begin{aligned} a_{N_1} &= \Delta H_{N_1} + \tfrac{1}{2} V_{xx}^{N_1+1}(\bar{x}_{N_1+1})(\Delta f_{N_1})^2 \\ &\doteqdot -\varepsilon (H_u^{N_1})^T (H_u^{N_1}) \end{aligned}$$

or

$$u^*_{N_1} = u_{N_1} - \varepsilon C_{N_1}^{-1} H_u^{N_1}, \qquad 0 < \varepsilon \leqslant 1$$

whence

$$a_{N_1} \doteqdot -\varepsilon(1-\varepsilon/2)(H_u^{N_1})^T C_{N_1}^{-1} (H_u^{N_1})$$

where now the unspecified arguments are $\bar{x}_i$, $\bar{u}_i$, and $V_x^{i+1}(\bar{x}_{i+1})$; ε is now chosen so that the acceptability criterion is satisfied.

Application of the above procedure ensures finally that $|u_i - u_i^*|$ is sufficiently small, $i = 0, \ldots, N-1$, to guarantee an improved cost. If $N_1 = N_{\text{eff}}$, then $u_i = \bar{u}_i$, $i = 0, \ldots, N_1 - 1$, $u_{N_1} = u^*_{N_1}$ is given by the above procedure, thus ensuring $u^*_{N_1} - \bar{u}_{N_1}$ is of order ε, and u_i, $i > N_1$, is given by

$$x_{i+1} = f_i(x_i, u_i), \qquad x_{N_1} = \bar{x}_{N_1}$$

$$u_{N_1} = u^*_{N_1}$$

$$u_i = \bar{u}_i + \beta_i(x_i - \bar{x}_i), \qquad i > N_1$$

In this case $u_i^* - \bar{u}_i = 0$, $i \neq N_1$, and $u^*_{N_1} - \bar{u}_{N_1} = 0(\varepsilon)$.

The second-order optimization algorithm. The algorithm is essentially the same as that described in Section 4.2, except that the step size is adjusted by the procedure outlined above.

The algorithm also has one-step convergence when applied to the LQP problem. Though the order of the error of the estimate is the same as that of the algorithm of Section 4.2, the error terms e_i', e_i'', e_i''' are smaller for a given change $u_i^* - \bar{u}_i(\alpha_i)$, since many terms are now evaluated at $(\bar{x}_i, u_i^*)$ instead of $(\bar{x}_i, \bar{u}_i)$. Hence larger changes of control should be possible, especially if the system is a discretization of a continuous-time system.

The control u_i^* is chosen to minimize Equation (4.4.18), so that

$$\text{(1)} \qquad \Delta H_i + \tfrac{1}{2} V_{xx}^{i+1} \Delta f_i^2 \leqslant 0$$

Hence

$$\text{(2)} \qquad H_u^i + V_{xx}^{i+1} \Delta f_1 f_u^i = 0$$

and

$$\text{(3)} \qquad H_{uu}^i + V_{xx}^{i+1} \Delta f_i f_{uu}^i > 0$$

at $u_i = u^*$, if (1) has a unique minimum. The fact that (1) is negative ensures that a_0, the estimated change in cost is negative, and, thus, the existence of an ε such that the actual change of cost is negative. Hence, the algorithm produces an improvement at each iteration.

4.4.2. The Vector Case

We repeat the discussion given in Section 4.4.1, introducing a few more terms to deal with the notational difficulties. Thus, Equation (4.4.1) of Section 4.4.1 becomes

$$\delta x_{i+1} = \Delta f_i + f_x^i \delta x_i + f_u^i \delta u_i + \tfrac{1}{2} w_i + y_i + \tfrac{1}{2} z_i \tag{4.4.25}$$

where w_i replaces $f_{xx}^i \delta x_i^2$ and is an n vector whose rth component is given by

$$(w_i)_r = \delta x_i^T [\partial^2 (f_i)_r / \partial x_i^2] \delta x_i \tag{4.4.26}$$

Similarly y_i replaces $f_{ux}^i \delta u_i \delta x_i$, and z_i replaces $f_{uu}^i \delta u_i^2$:

$$(y_i)_r = \delta x_i^T [\partial^2 (f_i)_r / \partial x_i \, \partial u_i] \delta u_i \tag{4.4.27}$$

$$(z_i)_r = \delta u_i^T [\partial^2 (f_i)_r / \partial u_i^2] \delta u_i \tag{4.4.28}$$

The unspecified arguments in Equations (4.4.25) to (4.4.28) are $\bar{x}_i$ and u_i^*; Δf_i is defined as before:

$$\Delta f_i = f_i(\bar{x}_i, u_i^*) - f_i(\bar{x}_i, \bar{u}_i) \tag{4.4.29}$$

The vector equivalents of most of the terms in Equation (4.4.10) are easily obtained†:

$$V_{xx}^{i+1}(\Delta f_i)^2 \to \Delta f_i^T V_{xx}^{i+1} \Delta f_i$$

$$H_x{}^i \delta x_i \to (H_x{}^i)^T \delta x_i, \qquad H_u{}^i \delta u_i \to (H_u{}^i)^T \delta u_i$$

$$V_{xx}^{i+1} \Delta f_i f_x{}^i \delta x_i \to \Delta f_i^T V_{xx}^{i+1} f_x{}^i \delta x_i$$

$$V_{xx}^{i+1} \Delta f_i f_u{}^i \delta u_i \to \Delta f_i^T V_{xx}^{i+1} f_u{}^i \delta u_i$$

$$H_{xx}^i \delta x_i^2 \to \delta x_i^T H_{xx}^i \delta x_i \qquad H_{uu}^i \delta u_i^2 \to \delta u_i^T H_{uu}^i \delta u_i$$

$$V_{xx}^{i+1}(f_x{}^i)^2 \delta x_i^2 \to \delta x_i^T (f_x{}^i)^T V_{xx}^{i+1} f_x{}^i \delta x_i$$

$$V_{xx}^{i+1}(f_u{}^i)^2 \delta u_i^2 \to \delta u_i^T (f_u{}^i)^T V_{xx}^{i+1} f_u{}^i \delta u_i$$

$$V_{xx}^{i+1} f_u{}^i f_x{}^i \delta u_i \delta x_i \to \delta u_i^T (f_u{}^i)^T V_{xx}^{i+1} f_x{}^i \delta x_i$$

$$V_{xx}^{i+1} \Delta f_i f_{xx}^i \delta x_i^2 \to \Delta f_i^T V_{xx}^{i+1} w_i$$

$$V_{xx}^{i+1} \Delta f_i f_{ux}^i \delta u_i \delta x_i \to \Delta f_i^T V_{xx}^{i+1} y_i$$

$$V_{xx}^{i+1} \Delta f_i f_{uu}^i \delta u_i^2 \to \Delta f_i^T V_{xx}^{i+1} z_i$$

The last three terms require further manipulation; $\Delta f_i^T V_{xx}^{i+1} w_i$ is a quadratic form in δx_i. Thus, there exists a symmetric matrix F_i such that

$$\delta x_i^T F_i \delta x_i = \Delta f_i^T V_{xx}^{i+1} w_i \tag{4.4.30}$$

Similarly, there exists a matrix G_i and a symmetric matrix H_i such that

$$\delta u_i^T G_i \delta x_i = \Delta f_i^T V_{xx}^{i+1} y_i \tag{4.4.31}$$

$$\delta u_i^T H_i \delta u_i = \Delta f_i^T V_{xx}^{i+1} z_i \tag{4.4.32}$$

† The argument of V_{xx}^{i+1} is $\bar{x}_{i+1}$, and the arguments of the remaining terms are $\bar{x}_i$, u_i^*, $V_x^{i+1}(\bar{x}_{i+1})$.

We now define the matrices A_i, B_i, and C_i as follows:

$$A_i = H_{xx}^i + (f_x^i)^T V_{xx}^{i+1} f_x^i + F_i \tag{4.4.33}$$

$$B_i = H_{ux}^i + (f_u^i)^T V_{xx}^{i+1} f_x^i + G_i \tag{4.4.34}$$

$$C_i = H_{uu}^i + (f_u^i)^T V_{xx}^{i+1} f_u^i + H_i \tag{4.4.35}$$

With this notation the vector equivalent of Equation (4.4.15) is

$$\begin{aligned} a_i + (V_x^i)^T \delta x_i + \tfrac{1}{2} \delta x_i^T V_{xx}^i \delta x_i = {} & a_{i+1} + \Delta H_i + \tfrac{1}{2} \Delta f_i^T V_{xx}^{i+1} \Delta f_i \\ & + \delta x_i^T [H_x^i + (f_x^i)^T V_{xx}^{i+1} \Delta f_i] \\ & + \delta u_i^T [H_u^i + (f_u^i)^T V_{xx}^{i+1} \Delta f_i] \\ & + \tfrac{1}{2} \delta x_i^T A_i \delta x_i + \delta u_i^T B_i \delta x_i \\ & + \tfrac{1}{2} \delta u_i^T C_i \delta u_i + e_i \end{aligned} \tag{4.4.36}$$

(The unspecified argument of V_x^{i+1} and V_{xx}^{i+1} is $\bar{x}_{i+1}$; the remaining unspecified arguments are $\bar{x}^i$, u_i^*, and $V_x^{i+1}(\bar{x}_{i+1})$.)

The vector equivalent of the various results given in Section 4.4.1 are now easily obtained. The parameters a_i, $V_x^i(\bar{x}_i)$, $V_{xx}^i(\bar{x}_i)$ that define the estimates of the local cost function due to the nonoptimal policy defined by

$$u_i = u_i^* + \delta u_i \tag{4.4.37}$$

$$\delta u_i = \beta_i \delta x_i \tag{4.4.38}$$

$i = 0, \ldots, N-1$ and u_i^* arbitrary,† are obtained by substituting Equations (4.4.37) and (4.4.38) into Equation (4.4.36) and equating coefficients, yielding:

$$\begin{aligned} a_i &= a_{i+1} + \Delta H_i + \tfrac{1}{2} \Delta f_i^T V_{xx}^{i+1} \Delta f_i \\ V_x^i &= H_x^i + \beta_i^T H_u^i + [f_x^i + f_u^i \beta_i]^T V_{xx}^{i+1} \Delta f_i \\ V_{xx}^i &= A_i + \beta_i^T B_i + B_i^T \beta_i + \beta_i^T C_i \beta_i \end{aligned} \tag{4.4.39}$$

---

† $|u_i^* - \bar{u}_i| = O(\varepsilon)$.

The solution of Equations (4.4.39) with boundary conditions:

$$a_N = 0$$

$$V_x^N = F_x(\bar{x}_N) \tag{4.4.40}$$

$$V_{xx}^N = F_{xx}(\bar{x}_N)$$

yields estimates with errors of $O(\varepsilon^3)$, $O(\varepsilon^2)$, and $O(\varepsilon)$, respectively.

The optimal local policy is obtained by first choosing u_i^* to minimize the right-hand side of Equation (4.4.36) with respect to u_i, with δu_i and δx_i zero:

$$u_i^* = u_i \qquad \text{such that} \quad H_i(\bar{x}_i, u_i, V_x^{i+1})$$

$$+ \tfrac{1}{2}[f_i(\bar{x}_i, u_i) - f_i(\bar{x}_i, \bar{u}_i)]^T V_{xx}^{i+1}[f_i(\bar{x}_i, u_i) - f_i(\bar{x}_i, \bar{u}_i)] \qquad \text{is minimum} \tag{4.4.41}$$

At $u_i = u_i^*$, we have

$$H_u^i + (f_u^i)^T V_{xx}^{i+1} \Delta f_i = 0 \tag{4.4.42}$$

Substituting u_i^* into Equation (4.4.36) and differentiating with respect to δu_i yields:

$$H_u^i + (f_u^i)^T V_{xx}^{i+1} \Delta f_i + B_i \delta x_i + C_i \delta u_i = 0 \tag{4.4.43}$$

Hence, if C_i is invertible, the optimal δu_i is given by

$$\delta u_i = -C_i^{-1} B_i \delta x_i \tag{4.4.44}$$

(The assumption that C_i is positive-definite, and therefore invertible, is reasonable because, from Equation (4.4.36), C_i, evaluated at u_i^*, is the second derivative, at $u_i = u_i^*$ with respect to u_i, of a function—Expression (4.4.41)—that achieves its minimum at $u_i = u_i^*$). With the optimum value of u_i^* and β_i, where $\beta_i = -C_i^{-1} B_i$, Equations (4.4.39) reduce to

$$a_i = a_{i+1} + \Delta H_i + \tfrac{1}{2} \Delta f_i^T V_{xx}^{i+1} \Delta f_i$$

$$V_x^i = H_x^i + (f_x^i)^T V_{xx}^{i+1} \Delta f_i, \qquad V_{xx}^i = A_i - \beta_i^T C_i \beta_i \tag{4.4.45}$$

[The argument of V_{xx}^{i+1} is $\bar{x}_{i+1}$; the remaining arguments are $\bar{x}_i$, u_i^*, and $V_x^{i+1}(\bar{x}_i)$.] The boundary conditions remain unaltered. The optimization algorithm is as described for the scalar case.

4.4.3. Terminal Constraints

Terminal constraints can be coped with by essentially the same techniques employed in Chapter 2. The appropriate difference equations for this case are given by Gershwin and Jacobson [2] along with full details of the successful application of the method to an optimal orbit transfer problem.

4.5. CONTINUOUS-TIME SYSTEMS

Consider a nonlinear continuous-time system defined by

$$\dot{x} = f(x, u; t), \qquad x(t_o) = x_o$$

$$V(x_o; t_o) = \int_{t_o}^{t_f} L(x, u; t)\, dt + F(x(t_f))$$

The appropriate differential equations for the various cases considered in this chapter are given below.

4.5.1. Nominal Control Function

Let the control function be $\bar{u}(t)$, $t_o \leqslant t \leqslant t_f$, and $\bar{x}(t)$ the corresponding state function. Let $\bar{V}(\bar{x}; t)$ denote the cost due to initial condition $x(t) = \bar{x}$ and control $\bar{u}(t)$. Then,

$$\bar{V}(\bar{x}; t) = \int_t^{t_f} L(\bar{x}, \bar{u}; \tau)\, d\tau + F(\bar{x}(t_f))$$

$$-\dot{\bar{V}}_x(\bar{x}; t) = H_x, \qquad \bar{V}_x(\bar{x}; t_f) = F_x(\bar{x}(t_f))$$

$$-\dot{\bar{V}}_{xx}(\bar{x}; t) = H_{xx} + f_x^T \bar{V}_{xx}(\bar{x}; t) + \bar{V}_{xx}(\bar{x}; t) f_x, \qquad \bar{V}_{xx}(\bar{x}; t_f) = F_{xx}(\bar{x}(t_f))$$

where the unspecified arguments are $\bar{x}$, $\bar{u}$, $\bar{V}_x(\bar{x}; t)$, t, and

$$H(x, u, V_x; t) = L(x, u; t) + V_x^T f(x, u; t).$$

4.5.2. Local Linear Policy—Small Variations in Control

The new policy is defined by

$$u(t) = \bar{u}(t) + \alpha(t) + \beta(t)\, \delta x(t)$$

$$\delta x(t) = x(t) - \bar{x}(t)$$

$$V(x;t) = \bar{V}(\bar{x};t) + a(t) + [V_x(\bar{x};t)]^T \delta x + \tfrac{1}{2}\delta x^T V_{xx}(\bar{x};t)\,\delta x + \cdots$$

Estimates of a, V_x, and V_{xx} are given by solutions of

$$-\dot{a} = H_u{}^T \alpha + \tfrac{1}{2}\alpha^T H_{uu}\alpha, \qquad a(t_f) = 0$$

$$-\dot{V}_x = H_x + \beta^T H_u + (H_{uu}\beta + B)^T \alpha, \qquad V_x(t_f) = F_x(\bar{x}(t_f))$$

$$-\dot{V}_{xx} = A + \beta^T H_{uu}\beta + \beta^T B + B^T \beta, \qquad V_{xx}(t_f) = F_{xx}(\bar{x}(t_f))$$

where the unspecified arguments are $\bar{x}$, $\bar{u}$, $V_x(\bar{x};t)$, and t, and

$$A = H_{xx} + V_{xx} f_x + f_x{}^T V_{xx}$$

$$B = H_{ux} + f_u{}^T V_{xx}$$

4.5.3. Local Optimal Linear Policy—Small Variations in Control

The local optimal linear policy is as given in Section 4.5.2, with

$$\alpha = -\varepsilon H_{uu}^{-1} H_u, \qquad \beta = -H_{uu}^{-1} B$$

yielding

$$-\dot{a} = -(\varepsilon - \varepsilon^2/2) H_u{}^T H_{uu}^{-1} H_u, \qquad -\dot{V}_x = H_x + \beta^T H_u$$

$$-\dot{V}_{xx} = A - \beta^T H_{uu}\beta$$

with the same boundary conditions. The unspecified arguments are $\bar{x}$, $\bar{u}$, $V_x(\bar{x};t)$, and t.

4.5.4. Local Linear Policy—Global Variations in Control

$$u(t) = u^*(t) + \beta(t)\,\delta x$$

yielding

$$-\dot{a} = \Delta H, \qquad -\dot{V}_x = H_x + \beta^T H_u + V_{xx}\Delta f$$

$$-\dot{V}_{xx} = A + \beta^T H_{uu}\beta + \beta^T B + B^T \beta$$

The unspecified arguments are $\bar{x}$, u^*, $V_x(\bar{x}; t)$, and t, and

$$\Delta H = H(\bar{x}, u^*, V_x; t) - H(\bar{x}, \bar{u}, V_x; t)$$

$$\Delta f = f(\bar{x}, u^*, V_x; t) - f(\bar{x}, \bar{u}, V_x; t)$$

4.5.5. Local Optimal Linear Policy—Global Variations in Control

The policy is as given in Section 4.5.4, with

$$u^* = u \qquad \text{such that} \quad H(\bar{x}, u, V_x; t) \quad \text{is minimum}$$

i.e.,

$$H_u = 0, \qquad u = u^*$$

Also,

$$\beta = -H_{uu}^{-1}(H_{ux} + f_u^T V_{xx}), \qquad u = u^*$$

yielding

$$-\dot{a} = \Delta H, \qquad -\dot{V}_x = H_x + V_{xx}\Delta f$$

$$-\dot{V}_{xx} = A - \beta^T H_{uu}\beta$$

The unspecified arguments are $\bar{x}$, u^*, $V_x(\bar{x}; t)$, and t.

4.6. SUMMARY

In this chapter we have presented two algorithms for optimizing discrete-time systems. The first essentially uses expansions about the points $(\bar{x}_i, \bar{u}_i)$, $i = 0, \ldots, N-1$, whereas the second (global variations in control) employs expansions about the points $(\bar{x}_i, u_i^*)$, $i = 0, \ldots, N-1$, where u_i^* is obtained from a minimizing procedure. Although the algorithms now require, naturally, the solution of difference equations instead of differential equations, they are, in other respects, slightly more complex than the equivalent continuous-time algorithms. The iterative equations for a_i, $V_x^{\,i}$ and V_{xx}^i involve more terms—e.g., H_{uu} is replaced by C_i—but also the global variation algorithm requires a refinement of the step-size adjustment procedure. We have also studied nonoptimal policies; this will be of help in the discussion of stochastic problems considered in the following chapters. A first-order algorithm employing global variations does not seem possible for the general nonlinear problem. It would require the minimization of H_i with respect to u_i, and this would not coincide with the u_i^* specified by Equation (4.4.42). This discrepancy does not occur in continuous-time problems.

References

1. D. Q. Mayne, *Intern. J. Control* **3**,85 (1966).
2. S. B. Gershwin and D. H. Jacobson, Harvard Univ. Tech. Rept, **TR 566** (August 1968).
3. E. Polak, in Proceedings of Second International Conference on Computing Methods in Optimization Problems, San Remo, Italy, 1968; published by Academic Press, 1969.
4. A. T. Fuller, *J. Electron. Control*, 17, 283 (1964).

Chapter 5

STOCHASTIC SYSTEMS WITH DISCRETE-VALUED DISTURBANCES

5.1. INTRODUCTION

5.1.1. The Stochastic Control Problem

In this chapter and the next we shall consider systems with stochastic disturbances. Our aim is to obtain algorithms, analogous to those presented in the previous chapters, for optimizing such systems. Our approach thus differs from that customarily adopted in the literature (e.g., [1-13]) where dynamic programming is used to determine the optimal cost function and the optimal control policy.

The problems that arise can be introduced by considering the conventional dynamic programming solution of the discrete-time stochastic control problem. The system is described by

$$x_{i+1} = f_i(x_i, u_i, w_i), \qquad i = 0, \ldots, N-1 \tag{5.1.1}$$

where x_i and u_i are the state and control variables and $\{w_i, \quad i = 0, \ldots, N-1\}$ is a sequence of independent random variables of known distributions $\mathscr{F}(w_i)$, $i = 0, \ldots, N-1$. The initial condition may be specified as an initial value of $x_o (x_o = \bar{x}_o)$ or as an initial distribution $\mathscr{F}(x_o)$ for x_o. The sequence $\{x_o, x_1, \ldots\}$ is a sequence of random variables and is called a process. Specifically, it is a Markov process because it satisfies (for given $\{u_i\}$):

$$\mathscr{F}(x_{i+1}|x_o, \ldots, x_i) = \mathscr{F}(x_{i+1}|x_i) \tag{5.1.2}$$

where $\mathscr{F}(\cdot)$ denotes a distribution, and $\mathscr{F}(\cdot|\cdot)$ a conditional distribution. From Equation (5.1.2) [if the necessary probability densities denoted by $p(\cdot)$ and $p(\cdot|\cdot)$ exist], application of Bayes' theorem, for given u_i, yields

$$p(x_{i+1}) = \int p(x_{i+1}|x_i)\, p(x_i)\, dx_i \tag{5.1.3}$$

Iteration of Equation (5.1.2) shows that $p(x_k)$, $k > r$ is determined solely by $p(x_r)$ and, of course, $u_r, \ldots, u_{k-1}$. This is analogous to the deterministic case, where x_k, $k > r$ is determined solely by x_r and $u_r, \cdots, u_{k-1}$; and justifies our use of x_i as the state of the systems at time i.

The basic sample space with elements ω is denoted by Ω. For this case, ω denotes $\{x_0, w_0, \dots, w_{N-1}\}$ (if x_0 is a random variable) and, hence, $\Omega = E_{n+Np}$, where p is the dimension of the vector w_i. If $x_0 = \bar{x}_0$, ω denotes $\{w_0, \dots, w_{N-1}\}$ and $\Omega = E_{Np}$.

The control may be specified either as an open-loop schedule of control actions $\{u_0, \dots, u_{N-1}\}$, denoted by U, or as a feedback policy π, which is a sequence of controls laws $\{h_0(\cdot), \dots, h_{N-1}(\cdot)\}$. The ith control law defines the control action as a function of x_i:

$$u_i = h_i(x_i), \qquad i = 0, \dots, N-1 \tag{5.1.4}$$

A schedule U may be regarded as a degenerate policy π, for which $h_i(x_i)$ does not depend on x_i, $i = 0, \dots, N-1$. To avoid needless repetition, therefore, π is sometimes used to denote a schedule U in the sequel.

Because of the random variables w_i, $i = 0, \dots, N-1$, and possibly x_0, selection of a particular policy π does not determine which trajectory $\{x_k\}$ occurs, as it does in the deterministic case, but only the probability of occurence (in the space of all possible trajectories) of each trajectory. More precisely, to each ω there is a corresponding trajectory or sequence of states, which may be denoted by $\{x_k(\omega)\}$, where $x_0(\omega), \dots, x_N(\omega)$ satisfy Equations (5.1.1) and (5.1.4), with x_0 and $w_1, \dots, w_{N-1}$ specified by ω. If $\pi = U$, the control actions $\{u_0, \dots, u_{N-1}\}$ are obviously predetermined; otherwise, the control actions are random variables, having values

$$u_i(\omega) = h_i(x_i(\omega)) \tag{5.1.5}$$

When policy π is adopted, the cost of particular trajectory with initial condition x_0 is defined to be

$$\tilde{V}_0(x_0, \pi, \omega) = \sum_{i=0}^{N-1} L_i(x_i, u_i) + F(x_N) \tag{5.1.6}$$

where L_i, $i = 0, \dots, N-1$, and F are nonnegative scalar functions, and $x_0, \dots, x_N$ satisfy Equation (5.1.1), with $u_i = h_i(x_i)$. This cost is a random variable, depending on $x_0, w_0, \dots, w_{N-1}$, and is thus a function of ω. The average cost of the process with initial condition x_0 due to a policy π is

$$V_0(x_0, \pi) = \underset{|x_0(\omega) = x_0}{E} [\tilde{V}_0(x_0(\omega), \pi, \omega)] \tag{5.1.7}$$

where $\underset{|x_0(\omega)=x_0}{E(\ )}$ denotes expectation with respect to the distributions of $x_1(\omega), \ldots, x_N(\omega)$ conditional on the random variable $x_0(\omega)$ having the value x_0.† Let $\phi(\pi)$ be defined as

$$\begin{aligned}\phi(\pi) &= EV_0(x_0(\omega), \pi) \\ &= E\tilde{V}_0(x_0(\omega), \pi, \omega)\end{aligned} \tag{5.1.8}$$

The stochastic problem to be considered may now be stated: "Find an admissible policy π (schedule U) such that $\phi(\pi)$ is minimized."

5.1.2. Determination of the Optimal Policy for Discrete-Time Systems

The techniquc of dynamic programming may be used to solve this problem. First, however, let us define a few useful terms. $\tilde{V}_i(x_i, \pi, \omega)$ denotes the cost of the random trajectory with initial condition x_i, policy π, and $w_i, \ldots, w_{N-1}$ specified by ω, i.e.,

$$\tilde{V}_i(x_i, \pi, \omega) = \sum_{k=i}^{N-1} L_i(x_i, u_i) + F(x_N) \tag{5.1.9}$$

where $x_i, \ldots, x_N$, and $u_i, \ldots, u_{N-1}$ satisfy Equations (5.1.10) and (5.1.11), with $w_i, \ldots, w_{N-1}$ specified by ω:

$$x_{k+1} = f_k(x_k, u_k, w_k) \tag{5.1.10}$$

$$u_k = h_k(x_k) \tag{5.1.11}$$

Clearly,

$$\tilde{V}_i(x_i, \pi, \omega) = L_i(x_i, u_i) + \tilde{V}_{i+1}(x_{i+1}, \pi, \omega) \tag{5.1.12}$$

The boundary condition for Equation (5.1.12)is

$$\tilde{V}_N(x_N, \pi, \omega) = F(x_N) \tag{5.1.13}$$

Iteration of Equation (5.1.12) with the boundary condition of Equation (5.1.13) yields the sequence of functions $\{\tilde{V}_0(x_0, \pi, \omega), \ldots, \tilde{V}_N(x_N, \pi, \omega)\}$.

† $\underset{|x_0(\omega)=x_0}{E(\ )}$ will sometimes be abbreviated to $\underset{|x_0}{E}$.

Similarly $V_i(x_i, \pi)$ denotes the *average* cost of the process with initial condition x_i and policy π, i.e.,

$$V_i(x_i, \pi) = \underset{|x_i}{E}\ \tilde{V}_i(x_i(\omega), \pi, \omega)$$

$$= \underset{|x_i}{E}\left[\sum_{k=i}^{N-1} L_k(x_k, h_k(x_k)) + F_N(x_N)\right] \tag{5.1.14}$$

Clearly,

$$V_i(x_i, \pi) = \underset{|x_i}{E}\left[L_i(x_i, h_i(x_i)) + V_{i+1}(x_{i+1}, \pi)\right] \tag{5.1.15}$$

where

$$x_{i+1} = f_i(x_i, h_i(x_i), w_i)$$

Hence, Equation (5.1.15) may be written :

$$V_i(x_i, \pi) = L_i(x_i, h_i(x_i)) + \int V_{i+1}(x_{i+1})\, d\mathscr{F}(x_{i+1}|x_i) \tag{5.1.16}$$

Also

$$V_N(x_N, \pi) = F(x_N) \tag{5.1.17}$$

where $\mathscr{F}(x_{i+1}|x_i)$ is the distribution of $x_{i+1}(\omega)$ conditional on the random variable $x_i(\omega)$ having the value x_i. Equation (5.1.16) if iterated, using the boundary condition of Equation (5.1.17), yields the sequence $\{V_i(x_i, \pi)\}$ of average cost functions; and, in particular, $V_o(x_o, \pi)$ the average cost of the process with initial condition x_o and policy π.

The above procedure requires little modification to yield the optimal cost functions and control laws. Let $h_i{}^o(\cdot)$ denote the optimal control law at time i, and $\pi^o = \{h_o{}^o(\cdot), \ldots, h_{N-1}^o(\cdot)\}$ the optimal policy. Let $V_i{}^o(x_i)$ denote $V_i(x_i, \pi^o)$ and assume that $V_{i+1}^o(\cdot), \ldots, V_N{}^o(\cdot)$ and $h_{i+1}^o(\cdot), \ldots, h_{N-1}^o(\cdot)$ have all been determined. Then the procedure of dynamic programming can be used to determine $V_i{}^o(\cdot)$ and $h_i{}^o(\cdot)$ by performing the following minimization for all $x_i \in E_N$:

$$V_i^o(x_i) = \min_{u_i} \underset{|x_i}{E}\left[L_i(x_i, u_i) + V_{i+1}^o(x_{i+1})\right] \tag{5.1.18}$$

where

$$x_{i+1} = f_i(x_i, u_i, w_i) \tag{5.1.19}$$

Equation (5.1.18) is sometimes referred to as the fundemantal recurrence relation of dynamic programming. Iteration of Equation (5.1.18) with the boundary condition of Equation (5.1.17) yields the sequences $\{V_0^o(\cdot), \ldots, V_N^o(\cdot)\}$ and $\{h_0^o(\cdot), \ldots, h_{N-1}(\cdot)\}$. The storage requirements are the same as for the equivalent deterministic problem, but of course the minimization is more difficult, requiring the evaluation of an expectation for each choice of u_i. Note that the conditional expectation automatically makes u_i a function of x_i.

5.1.3. Determination of the Optimal Policy for Continuous Time Systems

The continuous-time system is defined by

$$dx(t) = f(x, u; t)\,dt + G(t)\,dw(t) \tag{5.1.20}$$

where $w(t)$ is the Wiener process (which can be interpreted as the integral with respect to time of white noise). For our purpose, we can formally treat $dw(t)$ as a random variable with the following properties:

$$E[dw(t)] = 0 \tag{5.1.21}$$

$$\operatorname{var}[dw(t)] = E\,dw(t)\,dw^T(t) = I\,dt \tag{5.1.22}$$

$$E[dw(t_1)\,dw^T(t_2)] = 0, \qquad t_1 \neq t_2 \tag{5.1.23}$$

For convenience, let

$$G(t)\,G^T(t) = \sigma(t) \tag{5.1.24}$$

Then,

$$\operatorname{var}[G(t)\,dw(t)] = \sigma(t)\,dt \tag{5.1.25}$$

A policy π is defined by the control law

$$u(t) = h(x; t) \tag{5.1.26}$$

for all $x \in E_N$, $t \in [t_o, t_f]$. The basic sample space with elements ω is denoted by Ω. To each ω, there corresponds a trajectory $x(t)$, $t_o \leqslant t \leqslant t_f$ with initial condition $x(0, \omega) = x$, which satisfies Equation (5.1.20) with $u(t)$ specified by Equation (5.1.26). The cost of this trajectory is

$$\tilde{V}(x, \pi, \omega; t_o) = \int_{t_o}^{t_f} L(x, u; t)\, dt + F(x(t_f)) \tag{5.1.27}$$

[We shall sometimes denote $x(t)$ by $x(t, \omega)$ and $u(t)$ by $u(t, \omega)$ to denote the dependence on ω.] The cost of the portion of this trajectory over the interval $[t, t_f]$ is

$$\tilde{V}(x, \pi, \omega; t) = \int_{t}^{t_f} L(x, h(x; \tau); \tau)\, d\tau + F(x(t_f)) \tag{5.1.28}$$

where $x(t, \omega) = x$. The average cost of the process with initial condition x_o at t_o and policy π is

$$V(x, \pi; t_o) = \underset{|x(t_o) = x_o}{E} \tilde{V}(x(t_o), \pi, \omega; t_o) \tag{5.1.29}$$

Similarly, the average cost of the process with initial condition x at time t is

$$V(x, \pi; t) = \underset{|x(t) = x}{E} \tilde{V}(x(t), \pi, \omega; t) \tag{5.1.30}$$

From Equations (5.1.27) and (5.1.30), we obtain

$$V(x, \pi; t) = \underset{|x(t) = x}{E} [L(x(t), h(x(t); t); t)\,\delta t + V(x(t) + \delta x, \pi; t + \delta t) + \cdots]$$

where

$$\delta x = f(x(t), h(x(t); t); t)\,\delta t + G(t)\,\delta w + \cdots$$

Expanding the right-hand side in a Taylor series and performing the conditional expectation—neglecting terms of order higher than $0(\delta t)$, then allowing $\delta t \to 0$—yields finally

$$-(\partial/\partial t) V(x, \pi; t) = L(x, h(x; t); t) + [V_x(x, \pi; t)]^T f(x, h(x; t); t) + \tfrac{1}{2} \operatorname{trace} [\sigma(t) V_{xx}(x, \pi; t)] \tag{5.1.31}$$

The boundary condition for this partial differential equation is

$$V(x, \pi, t_f) = F(x) \tag{5.1.32}$$

Solution of Equation (5.1.31) yields $V(\cdot, \pi, \cdot)$.

Let π^o denote the optimal policy and $V^o(x; t)$ denote $V(x, \pi^o; t)$. The policy that minimizes $V(x, \pi; t_o)$ for all $x \in E_n$, is denoted by π^o and is specified by the control law:

$$u(t) = h^o(x; t) \tag{5.1.33}$$

for all $x \in E_n$, $t \in [t_o, t]$. $V^o(x, t)$ satisfies

$$-(\partial/\partial t) V^o(x, t) = \min_u [L(x, u; t) + [V_x{}^o(x; t)]^T f(x, u; t)$$
$$+ \tfrac{1}{2} \operatorname{trace} [\sigma(t) V_{xx}^o(x; t)]] \tag{5.1.34}$$

Performing the indicated minimization for all $x \in E_n$ yields $h^o(\cdot\,; t)$. Solution of Equation (5.1.34), with the boundary condition given by Equation (5.1.32), thus solves the optimal control problem.

5.1.4. Noisy Observations

In the preceding discussion we have assumed that the state x of the system is known. This is not always the case. The discrete-time system may be defined by

$$x_{i+1} = f_i(x_i, u_i, w_i), \qquad i = 0, \ldots, N-1 \tag{5.1.35}$$

$$y_i = m_i(x_i, v_i), \qquad i = 0, \ldots, N \tag{5.1.36}$$

where $\{w_i, i = 0, \ldots, N-1\}$ and $\{v_i, i = 0, \ldots, N\}$ are both sequences of independent random variables (w_i and v_i are not necessarily independent), and $\{y_i, i = 0, \ldots, N\}$ is a sequence of noisy observations. The control u_i is constrained to be a function of the available information $\{y_o, \ldots, y_i\}$, denoted I_i, and sometimes called the information state:

$$u_i = h_i(I_i) \tag{5.1.37}$$

Using the known distribution of (w_i, v_i), $i = 0, \ldots, N-1$, an iterative relation for determining the distribution $\mathscr{F}(x_i | I_i)$ of x_i conditional on the available information may be obtained. As before, let $V_i^\circ(\cdot)$, $i=0, \ldots, N-1$ denote the optimal cost functions. At time $N-1$, using the technique of dynamic programming, we have

$$V_{N-1}^\circ = \min_{u_{N-1}} \underset{|I_{N-1}}{E} [L_{N-1}(x_{N-1}, u_{N-1}) + F(x_N)] \qquad (5.1.38)$$

Since

$$\underset{|I_{N-1}}{E} F(x_N) = \underset{|I_{N-1}}{E} [\underset{|I_N}{E} F(x_N)] \qquad (5.1.39)$$

we can replace $F(x_N)$ in Equation (5.1.38) by $V_N{}^\circ(I_N)$, where $V_N{}^\circ(I_N)$ is defined by

$$V_N{}^\circ(I_N) = \underset{|I_N}{E} F(x_N) \qquad (5.1.40)$$

Performing the minimization in Equation (5.1.38), for all possible values of I_{N-1}, yields the optimal control law $h_{N-1}^\circ(I_{N-1})$ and the optimal cost function $V_{N-1}^\circ(I_{N-1})$ as functions of the information state I_{N-1}. The general recurrence relation is

$$V_i^\circ(I_i) = \min_{u_i} \underset{|I_i}{E} [L_i(x_i, u_i) + V_{i+1}^\circ(I_{i+1})] \qquad (5.1.41)$$

with the boundary condition given by Equation (5.1.40). Iteration of Equation (5.1.41) with this boundary condition effectively solves the optimal control problem but requires, at each stage, the computation of the conditional distribution $\mathscr{F}(x_i|I_i)$ and $\mathscr{F}(I_{i+1}|I_i)$. In some cases, and, in particular, the case when the system and observation equations are linear and the disturbances Gaussian, it becomes possible to obtain a sufficient statistic z_i for $\mathscr{F}(x_i|I_i)$. I_i can then be replaced by z_i, and $V_i{}^\circ(\cdot)$ and $h_i{}^\circ(\cdot)$ become functions of z_i, yielding the recurrence relation

$$V_i^\circ(z_i) = \min_{u_i} \underset{|z_i}{E} [L_i(x_i, u_i) + V_{i+1}^\circ(z_{i+1})] \qquad (5.1.42)$$

with boundary condition

$$V_N{}^\circ(z_N) = \underset{|z_N}{E} F(x_N) \qquad (5.1.43)$$

The symbol z_i is sometimes referred to as the hyperstate. Iterative equations, based on Baye's theorem, for determining $\mathscr{F}(x_i|I_i)$ and, therefore, z_i, $i=0, \ldots, N$ can be derived.

5.1.5. Conclusion

We have given a necessarily brief introduction to the optimal control problem when the system is stochastic. The basic technique of dynamic programming is discussed in detail by Bellman [1, 3], Bellman and Dreyfus [2], and Dreyfus [4]. The derivation of the partial differential equation for the optimal cost function is given in Florentin [5], Kushner [6], and Wonham [7]. The solution to the optimal control problem when the system is linear and the cost is quadratic is particularly simple [5]; the control policy is identical to that obtained when the white-noise disturbance is ignored. The problem of noisy observations is discussed by Florentin [8] and Wonham [9]; the latter gives a complete and rigorous derivation of the solution when the system is linear, the cost quadratic, and the disturbance Gaussian. The iterative equations, for deriving the conditional distribution of the state for the discrete-time system, are derived in Aoki [10]. The extension of the results of Section 5.1.4 to the continuous-time problem (involving both the derivation of the differential equation that describes the evolution of the conditional probability density function, and the application of this to the optimal control problem) is given by Kushner [11]. Many of these problems have been considered independently by Stratonovich [12] and Feldbaum [13]. The book by Dreyfus [2] gives a very lucid introduction to the stochastic control problem.

5.2. SYSTEM DESCRIPTION

The continuous-time system that we shall consider in this section is subjected to a random disturbance that has discrete values $d_1, \ldots, d_J$. The disturbance $w(t)$ changes value only at a finite set of times $t_1, \ldots, t_K$ in the interval $[t_o, t_f]$. The system is thus defined by

$$\dot{x}(t) = f(x, u, w; t) \tag{5.2.1}$$

$$x(t_o) = \bar{x}_o \tag{5.2.2}$$

where $w(t)$ is constant (having a value $d_1, d_2, \ldots,$ or d_J) in each subinterval $[t_i, t_{i+1})$, $i = 0, \ldots, K$, (t_{K+1} denoting t_f). The transition of $w(t)$ from d_r to d_s at time t_i is defined by the stochastic matrix P_i with elements p^i_{rs}:

$$\text{prob}\,[w(t_i) = d_s | w(t_{i-1}) = d_r] = p^i_{rs} \tag{5.2.3}$$

where

$$\sum_{s=1}^{J} p^i_{rs} = 1 \tag{5.2.4}$$

An initial condition must be assigned to $w(t_o)$ [which is the value of $w(t)$ in the interval $[t_o, t_1)$]. This can be specified as either:

$$w(t_o) = \bar{w}_o = d_1 \tag{5.2.5}$$

or

$$\text{prob}\,[w(t_o) = d_j] = p_j, \qquad j = 1, \ldots, J \tag{5.2.6}$$

where

$$\sum_{j=1}^{J} p_j = 1 \tag{5.2.7}$$

For simplicity, we shall employ the initial condition of Equation (5.2.5).

We also assume that the value of the disturbance is known to the controller instantaneously. The effective state of the process (since the disturbance is obviously a Markov process) is (x, w), and the optimal feedback control law is a function of (x, w, t).

The discrete-time system is similarly defined:

$$x_{i+1} = f_i(x_i, u_i, w_i) \tag{5.2.8}$$

$$x_o = \bar{x}_o, \qquad w_o = \bar{w}_o \tag{5.2.9}$$

The disturbance w_i is defined as before, the only extra requirement being that the times $t_1, \ldots, t_K$ belong to the set $\{0, T, 2T, \ldots, NT\}$, where T is the sampling interval of the discrete-time system, and $t_o = 0$, $t_{K+1} = t_f = NT$.

The free response of either the continuous-time or discrete-time system to a control schedule or control policy has the following properties. If the initial condition of the disturbance is specified by Equation (5.2.5), there is one trajectory in the interval $[t_o, t_1)$. This trajectory decomposes into J trajectories at time t_1, the rth trajectory having probability p_{1r}^1, $r = 1, \ldots, J$, where

$$\sum_{r=1}^{J} p_{1r}^1 = 1$$

At time t_2, *each* of the J trajectories decomposes into a further J trajectories. Thus, the rth trajectory [*i.e.*, the trajectory in the interval $[t_o, t_1)$ corresponding to $w(t) = d_r$] decomposes into J trajectories at time t_2 with probabilities, conditional on $w(t) = d_r$, $t \in [t_o, t_1)$, p_{rs}^2, $s = 1, \ldots, J$. Similarly, there is a further decomposition at t_3 (see Figure 5.1). If feedback control is employed, the control function undergoes a similar decomposition, as shown in Figure 5.1. For open-loop control, there is, of course, no decomposition of the control function.

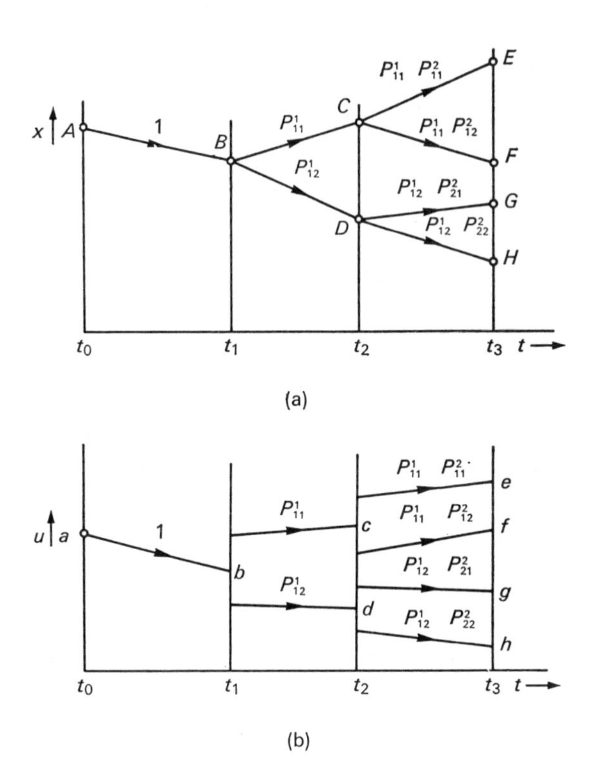

Figure 5.1

Let the state of the system be denoted by z:

$$z = (x, w) \tag{5.2.10}$$

The cost function is defined as before:

$$V(z; t_o) = \mathop{E}_{|z(t_o)=z} \left[\int_{t_o}^{t_f} L(x, u; t)\, dt + F(x(t_f)) \right] \tag{5.2.11}$$

for the continuous-time system, the dependence on the policy π being omitted for simplicity. For the discrete-time case,

$$V_o(z) = \underset{|z_o=z}{E} \left[\sum_{i=0}^{N-1} L_i(x_i, u_i) + F(x_N)\right] \tag{5.2.12}$$

The optimal control problem is to choose an admissible π to minimize $V(\bar{z}_o; t_o)$ or $V_o(\bar{z}_o)$, where $\bar{z}_o = (\bar{x}_o, d_1)$. With this specification of initial condition, the problem of determining a policy π reduces to that of finding a control function of the form shown in Figure 5.1 to minimize $V(\bar{z}_o; t_o)$ or $V_o(\bar{z}_o)$. This is very similar to the situation in the deterministic case where finding a policy π is replaced by the problem (if the initial condition is specified) of finding a control function $u(\cdot)$. The solution of this problem is aided by a simple addition rule developed in the next section.

5.3. THE ADDITION RULE

Consider a continuous-time system, described by Equation (5.2.1), with discrete disturbances and initial condition (x, w) at time t. The cost of each realization with initial condition $z = (x, w)$ at time t is

$$\tilde{V}(z, \omega; t) = \int_t^{t_f} L(x, u; \tau)\, d\tau + F(x(t_f)) \tag{5.3.1}$$

where, for $\tau \in [t, t_f]$, $x(\tau)$ satisfies Equation (5.2.1) with initial condition $x(t) = x$; $w(\tau)$ is specified by ω, $w(t) = w$; and $u(\tau)$ is specified by π [i.e., $u(\tau) = h(z(\tau); \tau)$].

For future convenience, let $\phi(t'|z, t)$; $z = (x, w)$ denote $(x(t'), w)$, where $x(t')$ is the solution of Equation (5.2.1) at time t', with disturbance $w(\tau) = w$, for all $\tau \in [t, t')$, $u(\tau)$ specified by π, and $x(t) = x$.

The average cost of all the realizations with initial condition $z(t) = z$ is

$$V(z; t) = \underset{|z(t)=z}{E} [\tilde{V}(z(t), \omega; t)] \tag{5.3.2}$$

At times t_i ($i = 1, \ldots, K$), w and, therefore, z suffer discontinuities. Given $z(t_i-)$, $z(t_i)$ and, hence, $V(z(t_i); t_i)$ are random variables. Thus,

$$V(z; t_i-) = \underset{|z(t_i-)=z}{E} [\tilde{V}(z(t_i), \omega; t_i)]$$

$$= \underset{|z(t_i-)=z}{E} \; \underset{|z(t_i)}{E} [\tilde{V}(z(t_i), \omega; t_i)]$$

i.e.,

$$V(z; t_i-) = \underset{|z(t_i-)=z}{E} [V(z(t_i); t_i)] \tag{5.3.3}$$

In terms of the stochastic matrix P_i, Equations (5.3.3) may be written:

$$V((x, d_r); t_i-) = \sum_{s=1}^{J} p_{rs}^i V((x, d_s); t_i), \qquad r = 1, \ldots, J \tag{5.3.4}$$

For a given initial condition (x, w) at time $t \in [t_{i-1}, t_i)$, there will be one state trajectory and one control function in the interval $[t, t_i)$ since the disturbance $w(\omega)$ will have only the one value w. Hence,

$$V(z; t) = \int_t^{t_i-} L(x, u; \tau)\, d\tau + V(z(t_i-); t_i-) \tag{5.3.5}$$

or

$$V((x, d_r); t) = \int_t^{t_i-} L(x, u; \tau)\, d\tau + V((x(t_i), d_r); t_i-) \tag{5.3.6}$$

where

$$z(\tau) = (x(\tau), w) = \phi(\tau | z, t) \tag{5.3.7}$$

for all $\tau \in [t, t_i)$.

Equations (5.3.3) and (5.3.4) form the addition rule. They show how, if the cost is obtained by integration in reverse time, the costs of different trajectories may be combined when trajectories coalesce. Together with Equations (5.3.5) and (5.3.6), they provide a means to calculate $V(z; t)$ that is preferable to averaging over all realizations [Equation (5.3.1)]. If we apply these equations to the example illustrated in Figure 5.1, where x_A denotes the value of x at point A, etc., we have

1. $$V((x_C, d_1); t_2) = \int_{CE} L\, d\tau + F(x_E)$$

2. $$V((x_C, d_2); t_2) = \int_{CF} L\, d\tau + F(x_F)$$

3. $$V((x_D, d_1); t_2) = \int_{DG} L\, d\tau + F(x_G)$$

4. $$V((x_D, d_2); t_2) = \int_{DH} L\, d\tau + F(x_H)$$

5. $$V((x_C, d_1); t_2\text{-}) = p^2_{11} V((x_C, d_1); t_2) + p^2_{12} V((x_C, d_2); t_2)$$

6. $$V((x_D, d_2); t_2\text{-}) = p^2_{21} V((x_D, d_1); t_2) + p^2_{22} V((x_D, d_2); t_2)$$

7. $$V((x_B, d_1); t_1) = \int_{BC} L\, d\tau + V((x_C, d_1); t_2\text{-})$$

8. $$V((x_B, d_2); t_1) = \int_{BD} L\, d\tau + V((x_D, d_2); t_2\text{-})$$

9. $$V((x_B, d_1); t_1\text{-}) = p^1_{11} V((x_B, d_1); t_1) + p^1_{12} V((x_B, d_2); t_1)$$

10. $$V((x_A, d_1); t_o) = \int_{AB} L\, d\tau + V((x_B, d_1); t_1\text{-})$$

On the other hand, application of Equation (5.3.1), i.e., integration along the basic realizations, yields

$$V((x_A, d_1); t_o) = p^1_{11} p^2_{11} [\int_{ABCE} L\, d\tau + F(x_E)]$$

$$+ p_{11}^1 p_{12}^2 \left[\int_{ABCF} L d\tau + F(x_F)\right]$$

$$+ p_{12}^1 p_{21}^2 \left[\int_{ABDG} L\, d\tau + F(x_G)\right]$$

$$+ p_{12}^1 p_{22}^2 \left[\int_{ABDH} L\, d\tau + F(x_H)\right]$$

Less integration is required using the first procedure. Let $\omega = \omega_1$ correspond to path ABCE, $\omega = \omega_2$ correspond to path ABCF, $\omega = \omega_3$ to path ABDG, and $\omega = \omega_4$ to path ABDH.

Because there is no additive white-noise term in the system equations, the cost function $V(z; t)$ satisfies, for $t \neq t_1, \ldots, t_K$:

$$-(\partial/\partial t) V((x, d_r); t) = L(x, u; t) + V_x^T((x, d_r); t) f(x, u, d_r; t) \quad (5.3.8)$$

where u is specified by π. Similarly, for $t \neq t_1, \ldots, t_K$, $\tilde{V}(z, \omega; t)$ satisfies

$$-(\partial/\partial t) \tilde{V}((x, d_r), \omega; t) = L(x, u; t) + \tilde{V}_x^T((x, d_r), \omega; t) f(x, u, d_r; t) \quad (5.3.9)$$

Equation (5.3.2) may be written

$$V((x, d_r); t) = \sum_{i \in I((x, d_r); t)} p_i \tilde{V}((x, d_r), \omega_i; t) \quad (5.3.10)$$

where $I(z; t) = \{i_1, \ldots, i_t\}$ is the index set specifying the set $\{\omega_{i_1}, \ldots, \omega_{i_t}\}$ of elements of Ω satisfying the initial condition $z(t) = (x, w)$, and $p_i = \text{prob}(\omega = \omega_i | z(t) = z)$, $i \in I(z; t)$, so that $\Sigma_{i \in I(z; t)} p_i = 1$. In our example (Figure 5.1), for $z = (x_C, d_1)$, $t = t_2-$, we have $I = (1, 2)$, $p_1 = p_{11}^2$, $p_2 = p_{12}^2$; for $z = (x_C, d_1)$, $t = t_2$, $I = (1)$, $p_1 = 1$, etc. Hence,

$$-\frac{\partial}{\partial t} V((x, d_r); t) = \sum_{i \in I((x, d_r); t)} p_i \left[-\frac{\partial}{\partial t} \tilde{V}((x, d_r), \omega_i; t)\right] \quad (5.3.11)$$

$$= \mathop{E}_{|z(t) = (x, d_r)} [L(x, u; t) + \tilde{V}_x^T((x, d_r), \omega; t) f(x, u, d_r; t)]$$

The optimal cost function satisfies

$$-\frac{\partial}{\partial t} V((x, d_r); t) = \min_u [L(x, u; t) + V_x^T((x, d_r); t) f(x, u, d_r; t)] \tag{5.3.12}$$

$$= \min_u \underset{|z(t)=(x,d_r)}{E} [L(x, u; t) + \tilde{V}_x^T((x, d_r), \omega; t) f(x, \dot{u}, d_r; t)]$$

Similar equations exist for the discrete-time system described by Equation (5.2.8). Thus, for any policy π, and $i+1 \neq t_1, \ldots, t_K$,

$$V_i(z_i) = L_i(x_i, u_i) + V_{i+1}(z_{i+1}) \tag{5.3.13}$$

where

$$z_{i+1} = f_i(x_i, u_i, w_i), \qquad z_i = (x_i, w_i)$$

For the optimal policy, $i+1 \neq t_1, \ldots, t_K$:

$$V_i(z_i) = \min_{u_i} [L_i(x_i, u_i) + V_{i+1}(z_{i+1})] \tag{5.3.14}$$

For $i+1 = t_i$, we have

$$V_{t_i-1}(z_{t_i-1}) = L_{t_i-1}(x_{t_i-1}, u_{t_i-1}) + V_{t_i-}(z_{t_i-}) \tag{5.3.15}$$

for policy π, and

$$V_{t_i-1}(z_{t_i-1}) = \min_{u_{t_i-1}} [L_{t_i-1}(x_{t_i-1}, u_{t_i-1}) + V_{t_i-}(z_{t_i-})] \tag{5.3.16}$$

for the optimal policy, where V_{t_i-} is defined by

$$V_{t_i-}(z_{t_i-}) = \underset{|z_{t_i-}}{E} [V_{t_i}(z_{t_i})] \tag{5.3.17}$$

i.e.,

$$V_{t_i-}(x_{t_i}, d_r) = \sum_{j=1}^{J} [p_{rs} V_{t_i}(x_{t_i}, d_s)] \tag{5.3.18}$$

for $r = 1, \ldots, J$.

5.4. DIFFERENTIAL DYNAMIC PROGRAMMING

The discussion in Section 5.3 is an application of dynamic programming, and enables the cost function $V(z; t)$ for all z and all $t \in [t_o, t_1]$ to be determined, for any given policy π. In this section we shall use differential dynamic programming to obtain an estimate of the cost $V(z; t)$ in the neighborhood of a family of nominal trajectories, and then employ these estimates in an optimization algorithm.

5.4.1. The Addition Rule

Assume that for a nominal policy $\bar{\pi}$, and initial condition $\bar{z}_o = (\bar{x}_o, \bar{w}_o)$ at time t_o, the resultant family of nominal state and control trajectories have been calculated using Equation (5.2.1). The resultant "family" will be similar to that shown in Figure 5.1. The nominal states at time t are denoted by $\bar{z}(t) = (\bar{x}(t), w(t))$, the disturbance $w(t)$ being unaffected by the choice of policy. At time t_i there are, say T_i values of $\bar{x}(t_i)$ (e.g., in Figure 5.1, $T_2 = 2$), and, for each value of $\bar{x}(t_i)$, there are J values $(d_1, \ldots, d_J)$ of $w(t_i)$. Hence there are JT_i values of $\bar{z}(t_i)$ at t_i, and this is the number of separate trajectories in the interval $[t_i, t_{i+1})$. Clearly,

$$T_{i+1} = JT_i$$

with initial condition

$$T_1 = 1$$

for the case $w(t_o) = d_1$ (also $T_o = 1$). The cost of the nominal policy $\bar{\pi}$ is denoted $\bar{V}(z; t)$, and we assume that $\bar{V}(\bar{z}(t); t)$ is known, $t \in [t_o, t_f]$.

Consider the effect of a new policy π. The cost of this policy is $V(z; t)$. Let the x neighborhood of $\bar{z} = (\bar{x}, w)$ denote the set of points $z = (x, w)$ for which $\|x - \bar{x}\|$ is sufficiently small, and assume that $V(z; t)$ can be approximated by a second-order Taylor series expansion about $\bar{z}(t)$ in a sufficiently small x neighborhood of $\bar{z}(t)$, $t \in [t_o, t_f]$:

$$\begin{aligned} V((x, d_r); t) = {} & \bar{V}((\bar{x}, d_r); t) + a((\bar{x}, d_r); t) + V_x^T((\bar{x}, d_r); t)\,\delta x \\ & + \tfrac{1}{2}\,\delta x^T V_{xx}((\bar{x}, d_r); t)\,\delta x + \cdots \end{aligned} \tag{5.4.1}$$

for $r = 1, \ldots, J$, where

$$\delta x = x - \bar{x} \tag{5.4.2}$$

Assume that the parameters a, V_x, V_{xx} are known for $t \in [t_i, t_f]$. From Equation (5.3.4),

$$\bar{V}((\bar{x}, d_r); t_i\text{-}) + a((\bar{x}, d_r); t_i\text{-}) + V_x^T((\bar{x}, d_r); t_i\text{-})\delta x + \tfrac{1}{2}\delta x^T V_{xx}((\bar{x}, d_r); t_i\text{-})\delta x$$

$$= \sum_{s=1}^{J} p_{rs}^i [\bar{V}((\bar{x}, d_s); t_i) + a((\bar{x}, d_s); t_i) + V_x^T((\bar{x}, d_s); t_i)\delta x$$

$$+ \tfrac{1}{2}\delta x^T V_{xx}((\bar{x}, d_s); t_i)\delta x] \tag{5.4.3}$$

But, from Equation (5.3.4),

$$\bar{V}((\bar{x}, d_r); t_i\text{-}) = \sum_{s=1}^{J} p_{rs}^i [\bar{V}((\bar{x}, d_s); t_i)] \tag{5.4.4}$$

Hence,

$$a((\bar{x}, d_r); t_i\text{-}) = \sum_{s=1}^{J} p_{rs}^i [a((\bar{x}, d_s); t_i)] \tag{5.4.5}$$

$$V_x((\bar{x}, d_r); t_i\text{-}) = \sum_{s=1}^{J} p_{rs}^i [V_x((\bar{x}, d_s); t_i)] \tag{5.4.6}$$

$$V_{xx}((\bar{x}, d_r); t_i\text{-}) = \sum_{s=1}^{J} p_{rs}^i [V_{xx}((\bar{x}, d_s); t_i)] \tag{5.4.7}$$

for $r = 1, \ldots, J$. Equations (5.4.1) and (5.4.5) to (5.4.7) may be written as

$$V(z; t) = \bar{V}(\bar{z}; t) + a(\bar{z}; t) + [V_x(\bar{z}; t)]^T \delta x + \tfrac{1}{2}\delta x^T [V_{xx}(\bar{z}; t)]\delta x + \cdots \tag{5.4.8}$$

$$a(\bar{z}; t_i\text{-}) = \underset{|z(t_i-)=\bar{z}}{E} [a(z(t_i); t_i)] \tag{5.4.9}$$

$$V_x(\bar{z}; t_i\text{-}) = \underset{|z(t_i-)=\bar{z}}{E} [V_x(z(t_i); t_i)] \tag{5.4.10}$$

$$V_{xx}(\bar{z}; t_i\text{-}) = \underset{|z(t_i-)=\bar{z}}{E} [V_{xx}(z(t_i); t_i)] \tag{5.4.11}$$

Equations (5.4.5) to (5. 4.7) [or (5.4.9) to (5.4.11)] form the addition rule of differential dynamic programming for this type of problem. From Equation (5.3.5), if $V(\bar{z}; t_i-)$ is known, $V(\bar{z}; t)$ for $t \in [t_{i-1}, t_i)$ can be obtained by integration along one trajectory. Hence if $a(\bar{z}; t_i-)$, $V_x(\bar{z}; t_i-)$, and $V_{xx}(\bar{z}; t_i-)$ are known, $a(\bar{z}; t)$, $V_x(\bar{z}; t)$, $V_{xx}(\bar{z}; t)$, $t \in [t_{i-1}, t_i)$ can be obtained by integration of the appropriate differential equations for a, V_x, and V_{xx}, given in Section 4.5, along each of the T_i trajectories, using $a(\bar{z}; t_i-)$, $V_x(\bar{z}; t_i-)$, $V_{xx}(\bar{z}; t_i-)$ as boundary conditions. Applying the addition rule again for each of the T_{i-1} values of $\bar{x}(t_{i-1})$ yields $a(\bar{z}; t_{i-1}-)$, $V_x(\bar{z}; t_{i-1}-)$, $V_{xx}(\bar{z}; t_{i-1}-)$, and the procedure can then be continued. The terminal conditions are:

$$a(\bar{z}; t_f) = 0 \tag{5.4.12}$$

$$V_x(\bar{z}; t_f) = F_x(\bar{x}(t_f)) \tag{5.4.13}$$

$$V_{xx}(\bar{z}; t_f) = F_{xx}(\bar{x}(t_f)) \tag{5.4.14}$$

for each of the T_f values of $\bar{x}(t_f)$. In Figure 5.1, for example, the addition rule would be applied at points C and D, and then again at B, as the integration proceeds in reverse time. The expected improvement in cost from replacing $\bar{\pi}$ by π is $a(\bar{z}_0; t_0)$.

It is also possible, of course, to integrate over the basic realizations. In the x neighborhood of $\bar{z}(t)$:†

$$\tilde{V}(z, \omega; t) = \bar{\tilde{V}}(\bar{z}, \omega; t) + \tilde{a}(\bar{z}, \omega; t) + \tilde{V}_x^T(\bar{z}, \omega; t)\,\delta x + \tfrac{1}{2}\delta x^T \tilde{V}_{xx}(\bar{z}, \omega; t)\,\delta x + \cdots \tag{5.4.15}$$

where the tilde denotes quantities referring to individual realizations (individual values of ω); $\tilde{a}$, $\tilde{V}_x$, $\tilde{V}_{xx}$ are obtained by integrating the appropriate differential equations given in Section 4.5 along each realization (for example, in Figure 5.1 this would mean integration along ECBA, FCBA, CDBA, and HDBA). From Equation (5.3.2),

$$a(\bar{z}; t) = \underset{|z(t)=\bar{z}}{E}[\tilde{a}(\bar{z}, \omega; t)] \tag{5.4.16}$$

$$V_x(\bar{z}; t) = \underset{|z(t)=\bar{z}}{E}[\tilde{V}_x(\bar{z}, \omega; t)] \tag{5.4.17}$$

$$V_{xx}(\bar{z}; t) = \underset{|z(t)=\bar{z}}{E}[\tilde{V}_{xx}(\bar{z}, \omega; t)] \tag{5.4.18}$$

† $\bar{\tilde{V}}$ is the cost of an individual realization using the nominal policy.

[for numerical evaluation, replace $E(\cdot)$ by

$$\sum_{i \in I(\bar{z};t)} p_i(\cdot)$$

see Equation (5.3.10)]. To use the differential equations of Section 4.5 either to determine $(\tilde{a}, \tilde{V}_x, \tilde{V}_{xx})$ or (a, V_x, V_{xx}), appropriate definitions of H must be employed. For the first case, we define

$$\tilde{H}(x, w, u, \lambda; t) = L(x, u; t) + \lambda^T f(x, u, w; t) \tag{5.4.19}$$

where

$$\lambda = \tilde{V}_x(z, \omega; t), \qquad w = w(\omega; t) \tag{5.4.20}$$

For the second case, we define

$$H(x, w, u, \lambda; t) = L(x, u; t) + \lambda^T f(x, u, w; t) \tag{5.4.21}$$

where now

$$\lambda = V_x(z; t) \tag{5.4.22}$$

From Equation (5.4.17), we have

$$H(x, w, u, V_x(z; t); t) = \underset{|z(t)=z}{E} [\tilde{H}(x, w, u, \tilde{V}_x(z, \omega; t); t] \tag{5.4.23}$$

since, given $z(t) = z$, only $\tilde{V}_x$ is a random variable.
Similar results hold for the discrete-time case. From Section 5.3,

$$\bar{V}_{t_i-}(x_{t_i}, d_r) = \sum_{s=1}^{J} p_{rs}^i \bar{V}_{t_i}(\bar{x}_{t_i}, d_s) \tag{5.4.24}$$

and:

$$a_{t_i-}(\bar{x}_{t_i}, d_r) = \sum_{s=1}^{J} p_{rs}^i a_{t_i}(\bar{x}_{t_i}, d_s) \tag{5.4.25}$$

$$V_x^{t_i-}(\bar{x}_{t_i}, d_r) = \sum_{s=1}^{J} p_{rs}^i V_x^{t_i}(\bar{x}_{t_i}, d_s) \tag{5.4.26}$$

$$V_{xx}^{t_i-}(\bar{x}_{t_i}, d_r) = \sum_{s=1}^{J} p_{rs}^i V_{xx}^{t_i}(\bar{x}_{t_i}, d_s) \tag{5.4.27}$$

for $r = 1, \ldots, J$.

5.4.2. A Second-Order Algorithm for the Determination of an Optimal (Feedback) Control Function

Feedback is used here in a restricted sense to denote that at time t the optimal $u^o(t)$ is required for each optimal state $(x^o(t), d_r)$, $r = 1, \ldots, J$, but not for all values of x, although the second-order algorithm yields an optimal policy for the x neighborhood of $(x^o(t), d_r)$, $r = 1, \ldots, J$. From Equation (5.3.17), the optimal value of $u(t)$ minimizes, with respect to u,

$$L(x, u; t) + V_x^T((x, d_r); t) f(x, u, d_r; t), \qquad r = 1, \ldots, J$$

Hence the natural extension of the second-order algorithm of Chapter 4 is:

1. For given nominal $\bar{u}(\omega; t)$, calculate $x(\omega; t)$ for all $t \in [t_o, t_f]$ and for all $\omega \in \Omega$ (all possible realizations of the disturbance sequence) using:

$$\dot{x} = f(x, u, w; t), \qquad x(t_o) = \bar{x}_o, \qquad w(t_o) = \bar{w}_o \tag{5.4.28}$$

2. Calculate, in reverse time, $a((\bar{x}, d_r); t)$, $V_x((\bar{x}, d_r); t)$, $V_{xx}((\bar{x}, d_r); t)$ (where $r = 1, \ldots, J$) for all $t \in [t_o, t_f]$ using:

$$-\dot{a} = \Delta H \tag{5.4.29}$$

$$-\dot{V}_x = H_x + V_{xx} \Delta f \tag{5.4.30}$$

$$-\dot{V}_{xx} = A - \beta^T H_{uu} \beta \tag{5.4.31}$$

(there are JT_i sets of these of equations in $[t_i, t_{i+1})$) with boundary conditions:

$$a((\bar{x}, d_r); t_f) = 0 \tag{5.4.32}$$

$$V_x((\bar{x}, d_r); t_f) = F_x(\bar{x}(t_f)) \tag{5.4.33}$$

$$V_{xx}((\bar{x}, d_r); t_f) = F_{xx}(\bar{x}(t_f)) \tag{5.4.34}$$

and matching conditions at $t_1, \ldots, t_K$:

$$a((\bar{x}, d_r); t_i-) = \sum_{s=1}^{J} p_{rs}^i \, a((\bar{x}, d_s); t_i) \tag{5.4.35}$$

$$V_x((\bar{x}, d_r); t_i-) = \sum_{s=1}^{J} p_{rs}^i \, V_x((\bar{x}, d_s); t_i) \tag{5.4.36}$$

$$V_{xx}((\bar{x}, d_r); t_i-) = \sum_{s=1}^{J} p_{rs}^i \, V_{xx}((\bar{x}, d_s); t_i) \tag{5.4.37}$$

[for all T_i values of $\bar{x}(t_i)$]. The unspecified arguments, in Equations (5.4.29) to (5.4.31) of a, V_x, V_{xx} are $((\bar{x}, d_r); t)$, and of the remaining variables $\bar{x}$, d_r, u_r^*, V_x, t. Also,

$$\Delta H = H(\bar{x}, d_r, u^*, V_x; t) - H(\bar{x}, d_r, \bar{u}, V_x; t) \tag{5.4.38}$$

$$\Delta f = f(\bar{x}, d_r, u^*; t) - f(\bar{x}, u, d_r; t) \tag{5.4.39}$$

$$u^* = \text{value of } u \text{ that minimizes } H(x, d_r, u, V_x; t) \tag{5.4.40}$$

$$\beta = -H_{uu}^{-1} B \tag{5.4.41}$$

$$A = H_{xx} + f_x^T V_{xx} + V_{xx} f_x \tag{5.4.42}$$

$$B = H_{ux} + f_u^T V_{xx} \tag{5.4.43}$$

(the unspecified arguments are, again, $\bar{x}$, d_r, u_r^*, V_x, t, etc.). Store $u^*(\bar{x}, d_r; t)$, $\beta(\bar{x}, d_r; t)$, $t \in [t_0, t_f]$, $r = 1, \ldots, J$. (Noting that at t_i- there are T_i values of u^* and β, and at t_i there are JT_i values.)

3. Generate new $x(\omega; t)$ and $u(\omega; t)$ using Equation (5.4.28) with

$$u(t) = \bar{u}(t), \qquad t \in [t_0, t_a] \tag{5.4.44}$$

$$u(t) = u^*(t) + \beta(t)[x(t) - \bar{x}(t)], \qquad t \in [t_a, t_f] \tag{5.4.45}$$

where the $u^*(t)$ and $\beta(t)$ appropriate to the original (nominal) state $(\bar{x}(t), d_r)$ are employed in Equations (5.4.44) and (5.4.45).

4. Increase t_a until the acceptability criteria (the change in $V_o(\bar{z}_o; t_o)$ is negative and greater in magnitude than $\frac{1}{2}|a(\bar{z}_o; t_o)|$) is satisfied.

5. Set $\bar{x}(\omega; t) = x(\omega; t)$, $\bar{u}(\omega; t) = u(\omega; t)$ for all $\omega \in \Omega$ and $t \in [t_o, t_f]$ and repeat Step 2.

6. Stop when $|a(\bar{z}_o; t_o)| \leqslant \eta$.

A similar algorithm can be constructed for the discrete-time case.

5.4.3. An Illustrative Example

$$x_{i+1} = x_i + u_i + w_i, \qquad x_o = 4$$

$$w_o = 0$$

$$\text{prob}\,[w_1 = \pm 1] = \tfrac{1}{2}$$

$$w_2 = 0$$

$$V_o = \underset{|x_o}{E}\left[\sum_{i=0}^{2} u_i^2/2 + x_3^2/2\right]$$

$$\bar{u}_o = \bar{u}_i = \bar{u}_2 = 0$$

There are only two possible disturbance sequences. The nominal state and control trajectories for the two cases ($w_1 = \pm 1$) are

	i	0	1	2	3
$\omega = \omega_1$	x_i	4	4	5	5
	w_i	0	1	0	—
	u_i	0	0	-0	—
$\omega = \omega_2$	x_i		4	3	3
	w_i		−1	0	—
	u_i		0	0	—

The optimal control will have one control action at $i = 0$ [for the single state (4, 0)], two at $i = 1$ [for the two states (x_1°, ± 1)] and two at $i = 2$ [for the two states (x_2°, 0], there being two values of x_2°]. Using the Equations (4.4.24)

(the discrete-time global variation algorithm), we obtain for this case

$$a_i = a_{i+1} + V_x^{i+1} u_i^* + \tfrac{1}{2}(1 + V_{xx}^{i+1})(u_i^*)^2$$

$$V_x^i = V_x^{i+1} + V_{xx}^{i+1} u_i^*$$

$$V_{xx}^i = V_{xx}^{i+1} - \beta_i^2(1 + V_{xx}^{i+1})$$

$$u_i^* \quad \text{minimizes} \quad u_i^2/2 + V_x^{i+1} u_i + \tfrac{1}{2} V_{xx}^{i+1} u_i^2$$

i.e.,

$$u_i^* = -V_x^{i+1}/(1 + V_{xx}^{i+1})$$

$$\beta_i = -V_{xx}^{i+1}/(1 + V_{xx}^{i+1})$$

Applying Step 2 of the algorithm (integrating a, V_x, and V_{xx} in reverse time, and applying the addition rule at $i = 1$) with boundary conditions:

$$a_3 = 0, \qquad V_x^3 = x_3 = 5 \text{ or } 3, \qquad V_{xx}^3 = 1$$

we obtain

i	0	1−	1	2	3
u_i^*	-1	—	$-\frac{5}{3}$	$-\frac{5}{2}$	—
β_i	$-\frac{1}{4}$	—	$-\frac{1}{3}$	$-\frac{1}{2}$	—
a_i	$-\frac{19}{3}$	$-\frac{17}{3}$	$-\frac{25}{3}$	$-\frac{25}{4}$	0
V_x^i	1	$\frac{4}{3}$	$\frac{5}{3}$	$\frac{5}{2}$	5
V_{xx}^i	$\frac{1}{4}$	$\frac{1}{3}$	$\frac{1}{3}$	$\frac{1}{2}$	1
u_i^*			-1	$-\frac{3}{2}$	—
β_i			$-\frac{1}{3}$	$-\frac{1}{2}$	—
a_i			-3	$-\frac{9}{4}$	0
V_x^i			1	$\frac{3}{2}$	3
V_{xx}^i			$\frac{1}{3}$	$\frac{1}{2}$	1

Applying Step 3 of the algorithm, the improved trajectories (with $t_2 = 0$) are

		i	0	1	2	3
ω_1		x_i	4	3	$\frac{8}{3}$	$\frac{4}{3}$
		w_i	0	1	0	—
		u_i	-1	$-\frac{4}{3}$	$-\frac{4}{3}$	—
ω_2		x_i		3	$\frac{4}{3}$	$\frac{2}{3}$
		w_i		-1	0	—
		u_i		$-\frac{2}{3}$	$-\frac{2}{3}$	—

The cost of the nominal trajectory is $\frac{17}{2}$, the expected reduction in cost, $|a_o|$ is $\frac{19}{3}$, yielding an estimated improved cost of $\frac{13}{6}$. The cost of the new trajectory is in fact $\frac{13}{6}$. This problem could be solved more simply, using initial nominal trajectories $\bar{x}_i = 0$, $i = 0, \ldots, 3$, and $\bar{u}_i = 0$, $i = 0, \ldots, 2$. An analogous simplification is discussed in the next section.

5.4.4. Linear Systems, Quadratic Cost Function

Consider the following system:

$$\dot{x}(t) = A(w; t)\, x(t) + B(w; t) u(t) \tag{5.4.46}$$

with cost function

$$V(z; t) = \underset{|z(t)=z}{E} \left[\int_t^{t_f} [\tfrac{1}{2} x^T Q x + \tfrac{1}{2} u^T R u]\, d\tau + \tfrac{1}{2} x^T(t_f) F x(t_f) \right] \tag{5.4.47}$$

where Q and F are positive semidefinite, R is positive-definite for all $\tau \in [t_o, t_f]$, $z = (x, w)$, and $w(t)$ is piecewise constant as before. For this type of problem, the differential dynamic programming procedure yields global results, and optimization is achieved in one iteration. Because of this, any nominal trajectory may be used, even one that does not satisfy the initial condition, and the simplest results are obtained for the nominal trajectories $\bar{x}(t) = 0$, $\bar{u}(t) = 0$ for $t \in [t_o, t_f]$. The parameters of the cost function are no longer dependent on x:

$$V((x, d_r); t) = a(d_r; t) + V_x^T(d_r; t) x + \tfrac{1}{2} x^T V_{xx}(d_r; t) x \tag{5.4.48}$$

for $r = 1, \ldots, J$. This has the enormous computional advantage of requiring backward integration of only J sets of differential equations for a, V_x, V_{xx} (in the general nonlinear case there are J^K sets in $[t_K, t_f]$, J^{K-1} in $[t_{K-1}, t_K)$— reducing to J values in $[t_1, t_2)$ and one value in $[t_o, t_1)$). Because of the symmetry of the problem, the term $V_x(d_r; t)$ is zero for all $t \in [t_o, t_f]$ and $r = 1, ..., J$; and since the cost of the nominal trajectory is zero, $a(d_r; t)$ is zero for all $t \in [t_o, t_f]$ and $r = 1, ..., J$. With these simplifications the algorithm becomes

1. Using the boundary conditions:

$$V_{xx}(d_r; t_f) = 0$$

integrate the following equations in backward time:

$$-\dot{V}_{xx}(d_r; t) = Q(t) + V_{xx}(d_r; t)A(d_r; t) + A^T(d_r; t)V_{xx}(d_r; t) - \beta^T(d_r; t)R(t)\beta(d_r; t)$$

where

$$\beta(d_r; t) = -[R(t)]^{-1}[B^T(d_r; t)V_{xx}(d_r; t)]$$

for $r = 1, ..., J$, employing the following matching conditions at $t_1, ..., t_K$:

$$V_{xx}(d_r; t_i\text{-}) = \sum_{s=1}^{J} p_{rs} V_{xx}(d_s; t_i).$$

2. The optimal control law is

$$u(x, d_r; t) = \beta(d_r; t)x(t).$$

The above results can be extended to deal with the case when $w(t)$ varies continuously with time, i.e.,

$$\begin{aligned}\text{prob}[w(t+\Delta) = d_s | w(t) = d_r] &= p_{rs}\Delta + O(\Delta), \qquad r \neq s \\ &= 1 + p_{rr}\Delta + O(\Delta), \qquad r = s\end{aligned}$$

The disturbance is now coupled into the differential equation for V_{xx} continuously. Thus,

$$\begin{aligned}V_{xx}(d_r; t) = V_{xx}(d_r; t+\Delta) &+ Q(t)\Delta + V_{xx}(d_r; t)A(d_r; t)\Delta \\ &+ A^T(d_r; t)V_{xx}(d_r; t)\Delta - \beta^T(d_r; t)R(t)\beta(d_r; t)\Delta \\ &+ \sum_{s=1}^{J} p_{rs} V_{xx}(d_s; t)\Delta + O(\Delta)\end{aligned}$$

where, as before,

$$\beta(d_r; t) = -[R(t)]^{-1}[B^T(d_r; t)V_{xx}(d_r; t)]$$

Allowing $\Delta \to 0$ yields

$$-\dot{V}_{xx}(d_r; t) = Q(t) + V_{xx}(d_r; t)A(d_r; t) + A^T(d_r; t)V_{xx}(d_r; t)$$
$$- \beta^T(d_r; t)R(t)\beta(d_r; t) + \sum_{s=1}^{J} p_{rs}V_{xx}(d_s; t)$$

which is the result given by Sworder [14] for this problem. The optimal control law is

$$u(x, d_r; t) = \beta(d_r; t)x(t)$$

5.4.5. Integration Along Basic Realizations

The algorithm described in Section 5.4.2 requires minimum integration. It is possible, however, to integrate $\tilde{a}$, $\tilde{V}_x$, and $\tilde{V}_{xx}$ along the basic realization (trajectories over the interval $[t_o, t_f]$ corresponding to realizations of ω), using the results of Section 5.4.1 and 5.4.2. However, the local control law at each time is not optimal for individual realizations (this would imply knowledge of the future disturbances). Therefore, even though we choose a locally optimal control law as we integrate in reverse time, we must use the nonoptimal differential equations (see Sections 4.5.2 and 4.5.4) for $\tilde{a}$, $\tilde{V}_x$, and $\tilde{V}_{xx}$. Using the definition of the stochastic Hamiltonian function [Equations (5.4.19) and (5.4.20)], the optimization algorithm becomes

1. Calculate the nominal state and control trajectories for each $\omega \in \Omega$.
2. For each realization (value of ω), integrate the following differential equations from t_f to t_o:

$$-\dot{\tilde{a}} = \Delta\tilde{H}$$

$$-\dot{\tilde{V}}_x = \tilde{H}_x + \beta^T\tilde{H}_u + \tilde{V}_{xx}\Delta f$$

$$-\dot{\tilde{V}}_{xx} = \tilde{A} + \beta^T\tilde{H}_{uu}\beta + \beta^T\tilde{B} + \tilde{B}^T\beta$$

where the unspecified arguments are $\bar{x}$, d_r, u^*, ω_i (specifying the realization), t, and

$$f = f(\bar{x}, u^*, d_r; t)$$

$$\Delta f = f(\bar{x}, u^*, d_r; t) - f(\bar{x}, \bar{u}, d_r; t)$$

$$\Delta\tilde{H} = \tilde{H}(\bar{x}, d_r, u^*, \lambda; t) - \tilde{H}(\bar{x}, d_r, \bar{u}, \lambda; t)$$

$$\lambda = \tilde{V}_x$$

$$u^* \text{ minimizes } [\underset{|x(t)=(\bar{x}, d_r)}{E} \tilde{H}(\bar{x}, d_r, u, \tilde{V}_x; t)]$$

$$= [\sum_{i \in I((\bar{x}, d_r); t)} p_i \tilde{H}(\bar{x}, d_r, u, \tilde{V}_x(\bar{x}, d_r, \omega_i; t); t)]$$

where $I(\bar{z}; t)$ is the index set specifying the values ω_i of ω corresponding to realizations and passing through the point $(\bar{x}, d_r)$ at time t; and where $p_i = \text{prob}\,[\omega = \omega_i | z(t) = \bar{z}]$ [$u^*(\bar{z}; t)$ is common to realizations ω_i, $i \in I(\bar{z}, t)$].

$$\tilde{A} = \tilde{H}_{xx} + \tilde{V}_{xx} f_x + f_x^T \tilde{V}_{xx}$$

$$\tilde{B} = \tilde{H}_{ux} + f_u^T \tilde{V}_{xx}$$

$$B = \underset{|z(t)=(\bar{x}, d_r)}{E} \tilde{B} = \sum_{i \in I((\bar{x}, d_r); t)} p_i \tilde{B}(\omega_i)$$

$$H_{uu} = \underset{|z(t)=(\bar{x}, d_r)}{E} \tilde{H}_{uu} = \sum_{i \in I((\bar{x}, d_r); t)} p_i \tilde{H}_{uu}(\omega_i)$$

$$\beta = -H_{uu}^{-1} B$$

where $\tilde{B}(\omega_i)$ denotes the value of $\tilde{B}$, at $((\bar{x}, d_r); t)$, corresponding to ω_i, etc. The boundary conditions are as in Section 5.4.2.

3. The rest of the algorithm is the same as that of Section 5.4.2.

5.4.6. First-Order Algorithm for the Determination of an Optimal (Feedback) Control Function

The algorithm is similar to that of Section 5.4.2, except that the differential equations for the parameters of the cost function are

$$-\dot{a} = \Delta H$$

$$-\dot{V}_x = H_x$$

with the same boundary conditions and matching conditions given previously. Here,

$$u^* = \text{value of } u \text{ that minimizes } H(\bar{x}, d_r, u, V_x; t)$$

The improved trajectory is generated using:

$$u(t) = \bar{u}(t), \qquad t \in [t_0, t_a]$$

$$u(t) = u^*(t), \qquad t \in [t_a, t_f]$$

for each nominal state $(\bar{x}(t), d_r)$ at time t.

For integration along basic realizations, we have

$$-\dot{\tilde{a}} = \Delta\tilde{H}, \qquad -\dot{\tilde{V}}_x = \tilde{H}_x$$

and u^* = value of u that minimizes

$$H = \underset{|z(t)=(\bar{x},\, d_r)}{E} \tilde{H}(\bar{x}, d_r, u, \tilde{V}_x; t) = \sum_{i \in I(\bar{z};\, t)} p_i \tilde{H}$$

where

$$\tilde{V}_x = \tilde{V}_x(\bar{x}, d_r, \omega_i; t)$$

5.4.7. Necessary Condition for Optimality

It follows from the preceding discussion that a necessary condition for the optimality of $u^*(t)$ is that $u^*(t)$ minimizes $H(x, w, u, V_x((x, w); t); t)$ with respect to u. More precisely, let the optimal policy be defined by

$$u(\tau) = h^\circ(x, w; \tau)$$

and let

$$\beta(x, w; \tau) = h_x{}^\circ(x, w; \tau)$$

Let $w(\omega_i; \tau)$, $\tau \in [t, t_f]$, denote the disturbance at time τ specified by ω_i, $i \in I(z; t)$ $[w(\omega_i; t) = w$ for all $i \in I(z; t)$, the index set specifying the realizations of ω satisfying the initial condition $z(t) = (x, w)]$. Let $x^\circ(\omega_i; \tau)$ denote the solution at time τ of

$$\dot{x}(\tau) = f(x(\tau), u^\circ(\omega_i; \tau), w(\omega_i; \tau); \tau)$$

given

$$x(t) = x, \qquad w(t) = w$$

where

$$u^\circ(\omega_i; \tau) = h^\circ(x^\circ(\omega_i; \tau), w(\omega_i; \tau); \tau)$$

Let $\tilde{\lambda}^\circ(\omega_i; t)$ denote the solution at time t of

$$-\dot{\tilde{\lambda}}(\tau) = \tilde{H}_x(x^\circ(\omega_i; \tau), w(\omega_i; \tau), u^\circ(\omega_i; \tau), \lambda(\tau); \tau)$$
$$+ \beta^T(x^\circ(\omega_i; \tau), w(\omega_i; \tau); \tau)\tilde{H}_u(x^\circ(\omega_i; \tau), w(\omega_i; \tau), u^\circ(\omega_i;\tau)\, \lambda(\tau);\tau)$$

$$\tilde{\lambda}(t_f) = F_x(x^\circ(\omega_i; t_f)$$

where

$$\tilde{H}(x, w, u, \tilde{\lambda}; t) = L(x, u; t) + \tilde{\lambda}^T f(x, u, w; t)$$

Then

$$\underset{|z(t)=z}{E}\ [\tilde{H}(x, w, u^{o}, \tilde{\lambda}^{o}(\omega_i; t); t)] \leqslant \underset{|z(t)=z}{E}\ [\tilde{H}(x, w, u, \tilde{\lambda}^{o}\ \omega_i; t); t)]$$

where

$$u^{o} = h^{o}(x, w; t)$$

This may be expressed as

$$\sum_{i \in I(z;t)} p_i \tilde{H}(x, w, u^{o}, \tilde{\lambda}^{o}(\omega_i; t); t) \leqslant \sum_{i \in I(z;t)} p_i \tilde{H}(x, w, u, \tilde{\lambda}^{o}(\omega_i; t); t)$$

or

$$H(x, w, u^{o}, \lambda^{o}; t) \leqslant H(x, w, u, \lambda^{o}; t)$$

where

$$\lambda^{o} = \underset{|z(t)=z}{E}\ \tilde{\lambda}^{o}(\omega_i; t) = \sum_{i \in I(z;t)} p_i \tilde{\lambda}^{o}(\omega_i; t)$$

and

$$H(x, w, u, \lambda^{o}; t) = L(x, u; t) + (\lambda^{o})^{T} f(x, u, w; t)$$

λ^{o} may alternatively be evaluated using the procedure of Section 4.5.2 (i.e., employing the addition rule when trajectories coalesce).

Note that the differential equation for $\tilde{\lambda}(\omega_i;\tau)$ involves the term $\beta^{T}\tilde{H}_u$. This results from the fact that $u^{o}(\omega_i; \tau)$ is not optimal for each realization passing through the point $x^{o}(\omega_i; \tau)$ (this would imply knowledge of the future disturbance).

However, since the differential equation for $\tilde{\lambda}^{o}$ is linear, and since

$$\underset{|z(\tau)=z^{o}}{E}\ \tilde{H}_u = \sum_{i \in I(z^{o};\tau)} p_i \tilde{H}_u = 0$$

where

$$z^{o} = (x^{o}(\omega_i; \tau), d_r)$$

for all $\tau \in [t_0, t]$ if the policy is optimal, the minimum principle still holds if $\tilde{\lambda}^{o}(\omega_i; t)$ is the solution of

$$-\dot{\tilde{\lambda}} = \tilde{H}_x$$

$$\tilde{\lambda}(t_f) = F_x(x^{o}(\omega_i; t))$$

which is the form of the minimum principle given by Kushner [15]. However, the first form is more useful, as noted by Sworder [14].

5.4.8. A First-Order Algorithm for the Determination of an Optimal Open-Loop Control Function (Schedule)

In this case the controller has no information concerning the state, and the control is thus a single function of time (or a single sequence of control actions in the discrete-time case). A nominal control $\bar{u}(t)$, $t_o \leqslant t \leqslant t_f$ is chosen, and the family of nominal state trajectories is generated using

$$\dot{x}(t) = f(x, u, w; t), \qquad x(t_o) = \bar{x}_o, \qquad w(t_o) = \bar{\omega}_o$$

$$u(t) = \bar{u}(t)$$

for all possible disturbance sequences $\{w(t_i)\}$. If now $\bar{u}(t)$ is altered to $u^*(t)$, $t_o \leqslant t \leqslant t_f$ then, from Sections 5.4.1, 5.4.2, the estimated change in cost is:

$$a(\bar{z}_o; t_o) = \int_{t_o}^{t_f} [\underset{|z(t_o)=\bar{z}_o}{E} \Delta\tilde{H}]\, dt = \int_{t_o}^{t_f} [\underset{|z(t_o)=\bar{z}_o}{E} \Delta H]\, dt$$

where ΔH and $\Delta\tilde{H}$ are defined respectively, as in Section 5.4.2 and Section 5.4.5. Hence, the algorithm becomes

1. Calculate $\bar{x}(\omega; t)$ as above for all $\omega \in \Omega$.
2. Integrate in reverse time a and V_x using the boundary condition:

$$a((\bar{x}, d_r); t_f) = 0$$

$$V_x((\bar{x}, d_r); t_f) = F_x(\bar{x}(t_f))$$

for all T_f values of $\bar{x}(t_f)$, and the matching conditions:

$$a((\bar{x}, d_r); t_i-) = \sum_{s=1}^{J} p^i_{rs}\, a((\bar{x}, d_s); t_i)$$

$$V_x((\bar{x}, d_r); t_i-) = \sum_{s=1}^{J} p^i_{rs}\, V_x((\bar{x}, d_s); t_i)$$

for all T_i values of $\bar{x}(t_i)$, $i = 1, \ldots, K$, and the following sets of differential equations [in the interval $[t_i, t_{i+1})$ there are JT_i sets]:

$$-\dot{a} = \Delta H$$

$$-\dot{V}_x = H_x$$

where ΔH is defined by Equation (5.4.38), the unspecified arguments being $\bar{x}$, d_r, u^*, V_x, t; and $u^*(t)$ is the value of u that minimizes

$$\underset{|z(t_o)=\bar{z}_o}{E} [H(\bar{x}, w, u, V_x; t)] = \sum_{j=1}^{JT_i} q_j H_j = \sum_{i \in I(\bar{z}_o; t_o)} p_i \tilde{H}_i$$

if $t \in [t_i, t_{i+1})$, where q_j denotes the probability, conditional on $z(t_o) = \bar{z}_o$ of the jth value of $\bar{z}(t)$, and H_j is the value of H evaluated at this point. (For example, in Figure 5.1 for $t \in [t_1, t_2)$, $q_1 = p_{11}^1$, $q_2 = p_{12}^1$; for $t \in [t_2, t_3]$, $q_1 = p_{11}^1 p_{12}^2$, $q_2 = p_{11}^1 p_{12}^2$, etc;)$\cdot p_i = \text{prob}(\omega = \omega_i)$, conditional on $z(t_o) = \bar{z}_o$, and $\tilde{H}_i$ is the value $\tilde{H}$ evaluated at a point on this realization.

3. Set

$$u(t) = \bar{u}(t), \qquad t \in [t_0, t_a)$$

$$u(t) = u^*(t), \qquad t \in [t_a, t_f]$$

and adjust t_a to satisfy an acceptance criterion

4. Set

$$\bar{u}(t) = u(t)$$

and repeat Step 1.

5. Stop when $|a(\bar{z}_o, t_o)| \leqslant \eta$.

An alternative procedure is to generate an improved control using

$$u(t) = \bar{u}(t) - \varepsilon \underset{|z(t_o) = \bar{z}_o}{E} [H_u]$$

and adjust ε until an acceptance criterion is satisfied. The latter procedure can also be employed for the discrete-time problem.

A full second-order algorithm would be difficult to implement, since the improvement algorithm for $u(t)$ at $t \in [t_i, t_{i+1})$ would be

$$u(t) = u^*(t) + \beta(t)[X(t) - \bar{X}(t)]$$

where $X(t)$ is a vector consisting of all JT_i values of $x(t)$. Consequently, the control would have to be expressed as a function of $X(t)$, increasing substantially the dimensionality of the problem.

5.5. CONCLUSION

It was the main purpose of this chapter to introduce the problem of stochastic control by the consideration of systems where the number of possible disturbance sequences (each disturbance acting for a finite time) is finite. This makes possible the determination of *all* realizations of the state and control trajectories for a given nominal policy. The optimization methods of the earlier chapters can then be applied, but the differential or difference equation of the parameters specifying the improved cost function, must be integrated backward in time along several (not one) state and control trajectories. Apart from this difference the resultant algorithms are very similar to those for deterministic systems. The basic first-order algorithm was first described by Mayne [16], and a second-order algorithm by Westcott *et al.* [17].

Although these results are interesting in their own right, our main purpose in examining the effect of discrete-valued stochastic disturbance is to provide an introduction to the problem of optimal control in the presence of continuous disturbances, which is the topic of the next chapter. However, the results of this chapter can also be applied to interesting problems arising in the design of chemical plant [18]. The plant consists of several stages, each stage described by a nonlinear equation relating the output state to the input state and control. For stages connected in cascade, the methods described in Chapter 4 may be used. For processes with "separating" or "combining" branches, the methods of this chapter may be used, the different branches being analogous to different trajectories due to disturbances. This is probably a more useful application of the algorithms described in this chapter.

References

1. R. Bellman, "Dynamic Programming," Princeton Univ. Press, Princeton, New Jersey, 1957.
2. R. Bellman and S. E. Dreyfus, "Applied dynamic programming," Princeton Univ. Press, Princeton, New Jersey, 1962.
3. R. Bellman, "Adaptive Control Processes," Princeton Univ. Press, Princeton, New Jersey, 1961.
4. S. E. Dreyfus, "Dynamic Programming and the Calculus of Variations," Academic Press, New York, 1965.
5. J. J. Florentin, *J. Electron. Control* **10**, 473 (1961).
6. H. J. Kushner, *IRE Trans. Auto. Control* AC-7, 120 (1962).
7. M. Wonham, *IEEE Intern. Conv. Record* **4**, 000 (1963).
8. J. J. Florentin, *J. Electron. Control* **13**, 263 (1962).
9. M. Wonham, Lecture Notes on Stochastic Control, Brown Univ., Providence, Rhode Island, 1967.
10. M. Aoki, "Optimization of Stochastic Systems," Academic Press, New York, 1967.
11. H. J. Kushner, *J. Math. Anal. and Appl.* **8**, 332 (1964).
12. R. L. Stratonovich, "Conditional Markov Processes and Their Application to the Theory of Optimal Control," American Elsevier, New York, 1968.
13. A. A. Feldbaum, "Optimal Control Systems," Academic Press, New York, 1965.
14. D. D. Sworder, Feedback Control of a Class of Linear Stochastic Systems, in JACC Proceedings, 1968.
15. H. J. Kushner, *J. Math. Anal. and Appl.* **11**, 78 (1965).
16. D. Q. Mayne, *Proc. IFAC Symp. Adaptive Control, Teddington, England, 1965.*
17. J. H. Westcott, D. Q. Mayne, G. F. Bryant, and S. K. Mitter, *Proc. 3rd IFAC Congr., London, 1966.*
18. L. T. Fan, C. S. Wang, "The Discrete Maximum Principle," Wiley, New York, 1964.

Chapter 6

STOCHASTIC SYSTEMS WITH CONTINUOUS DISTURBANCES

6.1. INTRODUCTION

In this chapter we discard the restriction that the disturbance be a discrete random variable. This means that the number of possible realizations is no longer finite; to cope with this difficulty, we generate only a finite number of realizations, using a random number generator with appropriate distribution. The consequence of this is that we only obtain estimates (in the statistical sense) of the various quantities that we require, the variance of the estimates being inversely proportional to the number of realizations. In Sections 6.2 and 6.3 we show how the results of previous chapters may be applied to a random sample of realizations—naive Monte-Carlo—to yield estimates of the optimal control action for a given state or of an optimal (open-loop) control function. In section 6.4 we present briefly two methods of variance reduction and show how these may be applied to control problems in Sections 6.5 and 6.6.

The reader will note that we do not describe a procedure for the determination of an optimal (feedback) policy. The algorithm for the determination of an optimal control action for a given state could form the basis of such an algorithm, but this would be of doubtful interest. Even if an optimal policy could be determined, its implementation would, in most cases, be prohibitively difficult. In many cases a controller with a finite number of parameters is postulated, and one tries to find the optimal values of the parameters. The algorithms that are described are suitable for this problem, whether the parameters are required to be time-invariant or not. It is probably worth pointing out that dynamic programming, though very suitable for determining an optimal policy, is not generally useful for determining an optimal value of a parameter, or for determining an optimal schedule (open-loop control).

The algorithms presuppose that a model of the nonlinear system is available, which can be simulated on a digital computer. The disturbance process for the continuous-time system discussed in this chapter will not, therefore, be the nonphysical white-noise process, but some suitable approximation. The problem of approximating diffusion processes by physical processes, and vice versa, will not be discussed here (see, e.g., Clark [1] and Wong and Zakai [2]), nor will the even more difficult problem of whether the optimal control parameters of a physical process are an approximation to the optimal control parameters of the diffusion process, of which the physical process is an approximation. We confine ourselves to the derivation of optimization algorithms for systems already simulated on a digital computer. We assume also that all realizations of the optimal state and control trajectories are bounded.

6.2. "NAIVE" MONTE-CARLO GRADIENT TECHNIQUES FOR DISCRETE-TIME SYSTEMS

6.2.1. Determination of the Optimal Control Action for a Given State

The system is defined by†

$$x_{i+1} = f_i(x_i, u_i, w_i), \qquad i = 0, \ldots, N-1 \tag{6.2.1}$$

We assume that the control policy has been chosen for the interval $[k+1, N-1]$:

$$u_i = h_i(x_i), \qquad i = k+1, \ldots, N-1 \tag{6.2.2}$$

and consider the problem of determining u_k to minimize the cost of the process with initial condition $\bar{x}_k$ at time k:

$$V_k(\bar{x}_k) = \underset{|\bar{x}_k}{E} \left[\sum_{i=k}^{N-1} L_i(x_i(\omega), u_i(\omega)) + F(x_N(\omega)\right] \tag{6.2.3}$$

where $\{x_i(\omega)\}$ and $\{u_i(\omega)\}$ denote solutions of Equations (6.2.1) and (6.2.2) with initial state $\bar{x}_k$ and initial control u_k, and $(|\bar{x}_k)$ denotes "given $x_k(\omega) = \bar{x}_k$." The bracketed term is the so of a single realization, denoted $\tilde{V}_k(\bar{x}_k, \omega)$, i.e.,

$$V_k(\bar{x}_k) = \underset{|\bar{x}_k}{E} (\tilde{V}_k(x_k(\omega), \omega)) \tag{6.2.4}$$

Let $\{\bar{x}_i(\omega)\}$ and $\{\bar{u}_i(\omega)\}$ denote the solutions of Equations(6.2.1) and (6.2.2) with initial state $\bar{x}_k$ and initial control $\bar{u}_k$. Let the appropriate costs be $\bar{V}_k(\bar{x}_k)$ and $\tilde{\bar{V}}_k(\bar{x}_k, \omega)$. For both cases, the policy over the interval $[k+1, \ldots, N-1]$ is the same, and so is the cost of a realization over this interval if the initial condition x_{k+1} is the same:

$$\tilde{V}_{k+1}(x_{k+1}, \omega) = \tilde{\bar{V}}_{k+1}(x_{k+1}, \omega) \tag{6.2.5}$$

† $\{w_i\}$ is a sequence of independent random variables.

From Section 4.2, the first and second derivatives, $\tilde{V}_x^{k+1}(\bar{x}_{k+1}(\omega), \omega)$ and $\tilde{V}_{xx}^{k+1}(\bar{x}_{k+1}(\omega), \omega)$, are obtained as solutions to the following equations:

$$\tilde{V}_x^{\,i} = (\tilde{H}_x^{\,i})' \tag{6.2.6}$$

$$\tilde{V}_{xx}^{i} = (\tilde{H}_{xx}^{i})' + (f_x^{\,T})' \tilde{V}_{xx}^{i+1} (f_x)' \tag{6.2.7}$$

with terminal conditions:

$$\tilde{V}_x^{\,N} = F_x(\bar{x}_N(\omega)) \tag{6.2.8}$$

$$\tilde{V}_{xx}^{N} = F_{xx}(\bar{x}_N(\omega)) \tag{6.2.9}$$

The unspecified arguments are $\bar{x}_i(\omega)$, $\bar{u}_i(\omega)$, $\tilde{V}_x^{i+1}$, and ω. $\tilde{H}'$ and f' are defined by

$$(\tilde{H}_i(x_i, \tilde{V}_x^{i+1}))' = \tilde{H}_i(x_i, h_i(x_i), \tilde{V}_x^{i+1}) \tag{6.2.10}$$

where

$$\tilde{H}_i(x_i, u_i, \tilde{V}_x^{i+1}) = L_i(x_i, u_i) + (\tilde{V}_x^{i+1})^T f_i(x_i, u_i, w_i) \tag{6.2.11}$$

$$(f_i(x_i, w_i))' = f_i(x_i, h_i(x_i), w_i) \tag{6.2.12}$$

Since Equations (4.3.12) and (4.3.13) hold for each realization, we have

$$\begin{aligned} \tilde{V}_k(\bar{x}_k + \delta x_k, \omega) = {} & \tilde{\tilde{V}}_k(\bar{x}_k, \omega) + [\tilde{H}_x^{\,k}]^T \delta x_k \\ & + [\tilde{H}_u^{\,k}]^T \delta u_k + \tfrac{1}{2} \delta x_k^{\,T} \tilde{A}_k \delta x_k \\ & + \delta u_k^T \tilde{B}_k \delta x_k + \tfrac{1}{2} \delta u_k^{\,T} \tilde{C}_k \delta u_k + \cdots \end{aligned} \tag{6.2.13}$$

where $\delta u_k = u_k - \bar{u}_k$
and $\tilde{A}_k$, $\tilde{B}_k$, and $\tilde{C}_k$ are defined as in Section 4.3.1:

$$\tilde{A}_k = \tilde{H}_{xx}^k + [f_x^{\,k}]^T \tilde{V}_{xx}^{k+1} f_x^{\,k} \tag{6.2.15}$$

$$\tilde{B}_k = \tilde{H}_{ux}^k + [f_u^{\,k}]^T \tilde{V}_{xx}^{k+1} f_x^{\,k} \tag{6.2.16}$$

$$\tilde{C}_k = \tilde{H}_{uu}^k + [f_u^{\,k}]^T \tilde{V}_{xx}^{k+1} f_u^{\,k} \tag{6.2.17}$$

The unspecified arguments in Equations (6.2.13) to (6.2.17) are $\bar{x}_k$, $\bar{u}_k$, $\tilde{V}_x^{k+1}$, and ω.

Taking the expectation of both sides of equation (6.2.13) conditional on $x_k(\omega) = \bar{x}_k$ yields

$$\begin{aligned} V_k(\bar{x}_k + \delta x_k) = {} & \bar{V}_k(\bar{x}_k) + [H_x{}^k]^T \delta x_k + [H_u{}^k]^T \delta u_k \\ & + \tfrac{1}{2} \delta x_k{}^T A_k \delta x_k + \delta u_k{}^T B_k \delta x_k \\ & + \tfrac{1}{2} \delta u_k{}^T C_k \delta u_k + \cdots \end{aligned} \tag{6.2.18}$$

where

$$\begin{aligned} H_x{}^k &= \underset{|\bar{x}_k}{E}\, \tilde{H}_x{}^k \\ H_u{}^k &= \underset{|\bar{x}_k}{E}\, \tilde{H}_u{}^k \\ A_k &= \underset{|\bar{x}_k}{E}\, \tilde{A}_k \\ B_k &= \underset{|\bar{x}_k}{E}\, \tilde{B}_k \\ C_k &= \underset{|\bar{x}_k}{E}\, \tilde{C}_k \end{aligned} \tag{6.2.19}$$

[The arguments of $\tilde{H}_x^k$, $\tilde{H}_u^k$, $\tilde{A}_k$, $\tilde{B}_k$, and $\tilde{C}_k$ are $\bar{x}_k$, $\bar{u}_k$, $\tilde{V}_x^{k+1}$, and $\tilde{V}_{xx}^{k+1}$; the arguments of $\tilde{V}_x^{k+1}$, $\tilde{V}_{xx}^{k+1}$ are $\bar{x}_{k+1}(\omega), \omega$).]

If the quantities in Equation (6.2.19) could be determined, by averaging over all realizations with initial condition $\bar{x}_k$, a suitable algorithm would be

1. For given $\bar{x}_k$ and $\bar{u}_k$, generate $\{\bar{x}_i(\omega)\}$ and $\{\bar{u}_i(\omega)\}$ for all ω using Equations (6.2.1) and (6.2.2).

2. Using Equations (6.2.6) to (6.2.9), determine $\tilde{V}_x^{k+1}(\bar{x}_{k+1}(\omega), \omega)$ and $\tilde{V}_{xx}^{k+1}(\bar{x}_{k+1}(\omega), \omega)$ for all ω. Calculate B_k, C_k, and $H_u{}^k$ using Equation (6.2.19).

3. Set $u_k = \bar{u}_k - \varepsilon C_k^{-1} H_u{}^k$, adjust ε until an acceptability criterion is satisfied. ($\Delta V_k = V_k(\bar{x}_k) - \bar{V}_k(\bar{x}_k)$ is negative and $|\Delta V_k|/|a_k| > 0.5$, where

$$a_k = -\varepsilon(1-\varepsilon/2)(H_u{}^k)^T C_k^{-1} H_u{}^k)$$

4. Set $\bar{u}_k = u_k$ and repeat Step 1.

5. Stop when $a_k \leqslant \eta$; $u_k{}^{\circ} = \bar{u}_k$.

The optimal control action for state $\bar{x}_k$ is denoted by $u_k{}^{\circ}$, and the optimal control action for state $\bar{x}_k + \delta x_k$, δx_k sufficiently small, is

$$u_k = u_k{}^{\circ} + \beta_k \delta x_k \tag{6.2.20}$$

$$\beta_k = -C_k^{-1} B_k \tag{6.2.21}$$

where C_k and B_k are evaluated as in Steps 1 and 2 using $\bar{u}_k = u_k{}^{\circ}$.

Naive Monte-Carlo calculation replaces the expectation in Equations (6.2.19) by a sample average over J realizations. Thus in Step 1 of the algorithms only J realizations ($\omega = \omega_1, \ldots, \omega_J$) of the disturbance sequence $\{w_k, \ldots, w_{N-1}\}$ are generated, using a random number generator, and hence only J realizations of $\{\bar{x}_i(\omega)\}$ and $\{\bar{u}_i(\omega)\}$. The expectations in Step 2 are replaced by averages, e.g., the estimate of H_u^k is

$$\hat{H}_u^{\,k} = (1/J) \sum_{j=1}^{J} \tilde{H}_u^{\,k}(\omega_j) \tag{6.2.22}$$

where $\tilde{H}_u^k(\omega_j)$ denotes the value of $\tilde{H}_u^k$ obtained using the jth realization of the disturbance sequence. The approximate variance of this estimate is

$$\text{var}(\hat{H}_u^{\,k}) \doteqdot (1/J^2) \sum_{j=1}^{J} [\tilde{H}_u^{\,k}(\omega_j) - \hat{H}_u^{\,k}][\tilde{H}_u^{\,k}(\omega_j) - \hat{H}_u^{\,k}]^T \tag{6.2.23}$$

and is of order $1/J$. Estimates of C_k and B_k may be similarly obtained. (Each realization has the same probability).

The efficient use of the estimates $\hat{H}_k$ and $\hat{C}_k$ to determine u_k° has not been fully resolved. While there is some relevant work on optimization using random gradients, there is little in the literature on the use of random second derivatives. A "brute force" method used with some success is to increase J until the standard deviation of $\delta u_k(-\varepsilon \hat{C}^{-1} \hat{H}_u^k)$ is small compared with δu_k. This automatically requires a larger J near the optimum ($H_u^k = 0$) until the required degree of precision is obtained.

The first order algorithm ($\delta u_k = -\varepsilon \hat{H}_u^k$) requires the backward integration of Equation (6.2.6) only, and does not yield an estimate of β_k for local optimal control. Methods of using the estimated gradient $\hat{H}_u^k$ have been discussed in the literature [3]. Usually, at the ith iteration of the algorithm, δu_k is set equal to $-\varepsilon_i \hat{H}_u^k$, where the sequence $\{\varepsilon_i\}$ satisfies

$$\sum_{i=1}^{\infty} \varepsilon_i = \infty, \qquad \sum_{i=1}^{\infty} \varepsilon_i^{\,2} < \infty$$

A sequence satisfying this is

$$\varepsilon_i = 1/i$$

Roughly speaking, the purpose of the decreasing value of ε_i is to ensure that when $u_k \to u_k^{\circ} (i \to \infty)$, $\delta u_k \to 0$ even though $\hat{H}_u^{\,k}$, being random, is nonzero. This requirement is not necessary in deterministic optimization, since $H_u^{\,k} \to 0$ as $u_k \to u_k^{\circ}$. However, the brute force method described above has proved satisfactory; by providing an estimate of the standard deviation of $\hat{H}_u^{\,k}$, it automatically provides a means for ascertaining J. J is relatively small when u_k differs appreciably from u_k°, and J increases as $u_k \to u_k^{\circ}$ until the required degree of precision is obtained.

For the second-order algorithm, the estimated change in cost

$$\hat{a}_o = -\varepsilon(1-\varepsilon/2)(\hat{H}_u{}^k)^T \hat{C}^{-1} \hat{H}_u{}^k$$

has a deterministic error of $0(\varepsilon^3)$ and a random error of $0(1/\sqrt{J})$. For the first-order algorithm, the estimated change in cost $\hat{a}_o = -\varepsilon(\hat{H}_u{}^k)^T \hat{H}_u{}^k$ has a deterministic error of $0(\varepsilon^2)$ and a random error of $0(1/\sqrt{J})$.

6.2.2. Determination of Optimal Parameters

Let z_i denote the state of the system at time i, and suppose the control law:

$$u_i = h_i'(z_i), \qquad i = 0, ..., N-1 \tag{6.2.24}$$

incorporates some time-invariant parameters θ that it is desired to optimize. Let θ_i, where

$$\theta_{i+1} = \theta_i, \qquad \theta_o = \bar{\theta}, \qquad i = 0, ..., N \tag{6.2.25}$$

be adjoined to form a new state vector $x_i = (z_i, \theta_i)$ satisfying the following difference equation:

$$x_{i+1} = f_i(x_i, u_i, w_i), \qquad x_o = \bar{x}_o \tag{6.2.26}$$

where $\bar{x}_o$ is, possibly, a random variable $\bar{x}_o(\omega)$; let the control law (6.2.24) be written in the form:

$$u_i = h_i(x_i), \quad i = 0, ..., N-1 \tag{6.2.27}$$

Then, for a given realization ω, the sequences $\{x_i(\omega)\}$ $\{u_i(\omega)\}$ may be determined, and hence (using Equations (6.2.6) and (6.2.7), $\tilde{V}_x{}^o(\bar{x}_o, \omega)$ $\tilde{V}_{xx}^o(\bar{x}_o, \omega)$. The appropriate components of $\tilde{V}_x{}^o$, $\tilde{V}_{xx}^o$ yield $\tilde{V}_\theta$ and $\tilde{V}_{\theta\theta}$ evaluated at $\theta = \theta$. By averaging over J realizations, the estimates $\hat{V}_\theta$ and $\hat{V}_{\theta\theta}$ of $E\tilde{V}_\theta$, $E\tilde{V}_{\theta\theta}$ may be obtained and employed to optimize θ using

$$\theta = \bar{\theta} - \varepsilon \hat{V}_{\theta\theta}^{-1} \hat{V}_\theta, \qquad 0 < \varepsilon \leqslant 1 \tag{6.2.28}$$

where ε is chosen to satisfy an acceptability criterion, and J to satisfy the requirement that the standard deviation of $\delta\theta$ is sufficiently smaller than $\delta\theta$. The estimated change in cost is $-\varepsilon(1-\varepsilon/2)\hat{V}_\theta{}^T \hat{V}_{\theta\theta}^{-1} \hat{V}_\theta$ and has a deterministic error of $0(\varepsilon^3)$ and a random error of $0(1/\sqrt{J})$.

In the first-order algorithm only $\tilde{V}_x{}^o(\bar{x}_o, \omega)$ is determined, using Equation (6.2.1), and the appropriate component $\tilde{V}_\theta$ of $\tilde{V}_x{}^o$ used to yield $\hat{V}_\theta$. The improvement rule is

$$\theta = \bar{\theta} - \varepsilon \hat{V}_\theta, \qquad \varepsilon > 0 \tag{6.2.29}$$

The estimated change in cost $-\varepsilon \hat{V}_\theta{}^T \hat{V}_\theta$, has a deterministic error $0(\varepsilon^2)$ and a random error $0(1/\sqrt{J})$.

6.2.3. Determination of Optimal Schedule

Let $\{x_i(\omega)\}$ denote the solution of Equation (6.2.1) with initial condition $x_0(\omega)$ and schedule $U=\{u_i\}$; and $\{\bar{x}_i(\omega)\}$ the solution with initial condition $x_0(\omega)$ and schedule $\bar{U}=\{\bar{u}_i\}$. The problem is the determination of U to minimize

$$\begin{aligned}\phi(U) &= E V_0(x_0(\omega)) \\ &= E \sum_{i=0}^{N-1} [L_i(x_i(\omega), u_i) + F(x_N(\omega))]\end{aligned} \tag{6.2.30}$$

U may specify a set of parameters of a time-varying control law. If

$$u_i = \bar{u}_i + \delta u_i, \qquad i = 0, \ldots, N-1 \tag{6.2.31}$$

where δu_i is of $0(\varepsilon)$, then, from Chapter 4, the estimated change of cost $\tilde{a}_0(\omega)$, of the realization ω, due to the adoption of U in place of $\bar{U}$, is the solution at time 0 of

$$\tilde{a}_i = \tilde{a}_{i+1} + (\tilde{H}_u^i)^T \delta u_i \tag{6.2.32}$$

$$\tilde{V}_x^i = \tilde{H}_x^i \tag{6.2.33}$$

with boundary conditions

$$\tilde{a}_N = 0 \tag{6.2.34}$$

$$\tilde{V}_x^N = F_x(\bar{x}_N(\omega)) \tag{6.2.35}$$

The unspecified arguments are $\bar{x}_i(\omega)$, $\bar{u}_i$, and $\tilde{V}_x^{i+1}$. This estimate has an error of $0(\varepsilon^2)$. Averaging over all realizations, we obtain

$$\phi(U) - \phi(\bar{U}) = a_0 + 0(\varepsilon^2) \tag{6.2.36}$$

where

$$a_0 = -\sum_{i=0}^{N-1} [E\tilde{H}_u^i]^T \delta u_i \tag{6.2.37}$$

If $E\,\tilde{H}_u^i$ were known, $i=0, \ldots, N-1$, we could set

$$\delta u_i = -\varepsilon H_u^i \tag{6.2.38}$$

where

$$H_u^i = E\tilde{H}_u^i \tag{6.2.39}$$

in which case

$$a_0 = -\varepsilon \sum_{i=0}^{N-1} (H_u^i)^T H_u^i \tag{6.2.40}$$

We could then adjust ε until an acceptability criterion is satisfied. In the naive Monte-Carlo algorithm, $H_u{}^i$ is replaced by

$$\hat{H}_u{}^i = (1/J) \sum_{j=1}^{J} \tilde{H}_u{}^i(\omega_j) \tag{6.2.41}$$

where $\tilde{H}_u{}^i(\omega_j)$ is the value of $\tilde{H}_u$ obtained using Equations (6.2.32) and (6.2.33) for the jth realization of $\{x_o(\omega), w_o(\omega), \ldots, w_{N-1}(\omega)\}$ which we denote $\omega = \omega_j$.

The acceptability criterion for the Monte-Carlo algorithm could be

$$|\hat{\phi}(U) - \hat{\phi}(\bar{U})|/|a_o| > 0.5 \tag{6.2.42}$$

where the averages, denoted by the caret, are done over the *same* J realizations $\omega_1, \ldots, \omega_J$ in order to avoid sampling errors in the difference $\hat{\phi}(U) - \hat{\phi}(U)$. A new set of J realizations would then be generated using the new schedule and a new realization of the disturbance sequence.

6.3. "NAIVE" MONTE-CARLO GRADIENT TECHNIQUES FOR CONTINUOUS-TIME SYSTEMS

6.3.1. Determination of the Optimal Control Action for a Given State

The system is defined by

$$\dot{x} = f(x, u; t) + w, \qquad t_o \leqslant t \leqslant t_f \tag{6.3.1}$$

where $w(t)$ is a "physical" random disturbance suitable for digital computer simulation. For example,

$$w(t) = w_i, \qquad i\Delta < t \leqslant (i+1)\Delta \tag{6.3.2}$$

where $\{w_i, i = 0, \ldots, N-1\}$ is a sequence of independent random variables of known distribution independent of x, and $\Delta = (t_f - t_o)/N$. The state of this system at time t is $(x(t), w(t))$, but we shall assume that the information available to the controller is $x(t)$. We assume that the policy π for the interval $(t+\varepsilon, t_f]$ has been chosen, and we consider the problem of choosing the control $u(\tau)$, $t \leqslant \tau \leqslant t+\varepsilon$, to minimize the cost of the process with initial condition

$$x(t) = \bar{x} \tag{6.3.3}$$

The policy π is defined by

$$u(\tau) = h(x; \tau), \qquad t+\varepsilon < \tau \leqslant t_f$$

For each realization $\omega \in \Omega$, there corresponds a particular disturbance $w(\omega; \tau)$, $t \leqslant \tau \leqslant t_f$; let $\bar{x}(\omega; \tau)$ and $\bar{u}(\omega; \tau)$ denote the corresponding solution of Equations (6.3.1) and (6.3.2) with initial conditions $x(t) = \bar{x}$, $u(t) = \bar{u}$;

and let $x(\omega; t)$, $u(\omega; t)$ denote the solutions with initial conditions $x(t) = \bar{x}$ and $u(t) = u^*$. In each case the control $u(\tau)$ is assumed to be constant (with value $\bar{u}$ or u^*) over the interval $[t, t+\varepsilon]$. We wish to choose u^* to minimize

$$V(\bar{x}; t) = \underset{|x(t)=\bar{x}}{E} [\tilde{V}(\bar{x}, \omega; t)] \tag{6.3.4}$$

where $\tilde{V}$ is the cost of an individual realization with $u(t) = u^*$.

To employ the results of Section 4.5, we need to define the Hamiltonian H. The obvious definition is

$$\tilde{H}(x, u, \tilde{V}_x; \tau) = L(x, u; \tau) + \tilde{V}_x{}^T[f(x, u; \tau) + w]$$

but because w does not depend on x and because we are only concerned with terms such as $\Delta\tilde{H}$, $\tilde{H}_x$, $\tilde{H}_u$, etc, where w does not appear, we may modify the definition of $\tilde{H}$ to

$$\tilde{H}(x, u, \tilde{V}_x; \tau) = L(x, u; \tau) + \tilde{V}_x{}^T f(x, u; \tau) \tag{6.3.5}$$

We will also need the following terms:

$$[\tilde{H}(x, \tilde{V}_x; \tau)]' = L(x, h(x; \tau); \tau) + \tilde{V}_x{}^T[f(x; \tau)]' \tag{6.3.6}$$

$$[f(x; \tau)]' = f(x, h(x; \tau); \tau) \tag{6.3.7}$$

If $\tilde{\tilde{V}}(\bar{x}, \omega; t)$ is the cost of a realization with $u(t) = \bar{u}$, then

$$\tilde{V}(\bar{x}, \omega; t) - \tilde{\tilde{V}}(\bar{x}, \omega; t) = \tilde{a}(\bar{x}, \omega; t) + o(\varepsilon^3)$$

$\tilde{a}(\bar{x}(\omega; \tau), \omega; \tau)$, $\tilde{V}_x(\bar{x}(\omega; \tau), \omega; \tau)$, and $\tilde{V}_{xx}(\bar{x}(\omega; \tau), \omega; \tau)$ are the solutions of

$$-\dot{\tilde{a}}(\tau) = 0$$
$$-\dot{\tilde{V}}_x(\tau) = \tilde{H}_x{}' \tag{6.3.8}$$
$$-\dot{\tilde{V}}_{xx}(\tau) = \tilde{H}'_{xx} + (f_x{}^T)' \tilde{V}_{xx}(\tau) f_x{}'$$

for $t+\varepsilon < \tau \leqslant t_f$. The terminal conditions are

$$\tilde{a}(t_f) = 0$$
$$\tilde{V}_x(t_f) = F_x(\bar{x}(\omega; t_f)) \tag{6.3.9}$$
$$\tilde{V}_{xx}(t_f) = F_{xx}(\bar{x}(\omega; t_f))$$

The unspecified arguments are $\bar{x}(\omega; \tau)$, $\bar{u}(\omega; \tau)$, $\tilde{V}_x(\tau)$ (and $\tilde{V}_x(\tau)$ is the value of $\tilde{V}_x(\bar{x}(\omega; \tau), \omega; \tau)$). The differential equation for $\tilde{a}(\tau)$ during the interval $[t, t+\varepsilon]$ is

$$-\dot{\tilde{a}}(\tau) = \Delta\tilde{H}(\tau) \tag{6.3.10}$$

Hence

$$\tilde{a}(t) = \Delta\tilde{H}(t)\varepsilon + 0(\varepsilon^2) \tag{6.3.11}$$

where

$$\Delta\tilde{H}(t) = \tilde{H}(\bar{x}, u^*, \tilde{V}_x(t); t) - \tilde{H}(\bar{x}, \bar{u}, \tilde{V}_x(t); t) \tag{6.3.12}$$

The optimal u^*, as $\varepsilon \to 0$, is the value of u that minimizes

$$\underset{|x(t)=\bar{x}}{E} \tilde{H}(\bar{x}, u, \tilde{V}_x(\bar{x}, \omega; t); t) \tag{6.3.13}$$

If we denote the estimate of $V(\bar{x}; t) - \bar{V}(\bar{x}; t)$ by $a(\bar{x}; t)$, i.e.,

$$a(\bar{x}; t) \doteqdot V(\bar{x}; t) - \bar{V}(\bar{x}; t) \tag{6.3.14}$$

where $\bar{V}$ is the average cost with $u(t) = \bar{u}$, then

$$a(\bar{x}; t) = \underset{|x(t)=\bar{x}}{E} \tilde{a}(\bar{x}, \omega; t) \tag{6.3.15}$$

Also,

$$V_x(\bar{x}; t) = \underset{|x(t)=\bar{x}}{E} \tilde{V}_x(\bar{x}, \omega; t) \tag{6.3.16}$$

$$V_{xx}(\bar{x}; t) = \underset{|x(t)=\bar{x}}{E} \tilde{V}_{xx}(\bar{x}, \omega; t) \tag{6.3.17}$$

Thus,

$$u^* = u \qquad \text{that minimizes} \quad H(\bar{x}, u, V_x(\bar{x}; t); t) \tag{6.3.18}$$

where

$$\begin{aligned} H(x, u, V_x; t) &= \underset{|x(t)=x}{E} \tilde{H}(x, u, \tilde{V}_x; t) \\ &= L(x, u, t) + V_x{}^T f(x, u, t) \end{aligned} \tag{6.3.19}$$

the arguments of V_x being $x; t$.

Consider now an initial state $\bar{x} + \delta x$. The optimal control $u^* + \delta u$ may be determined by expanding $\tilde{H}(x, u, \tilde{V}_x; t) + \tilde{V}_x{}^T w$ (the "original" $\tilde{H}$) up to terms of second-order in x and u about $(\bar{x}, u^*)$. $\Delta\tilde{H}(t)$ is replaced by

$$\begin{aligned} &\tilde{H}(\bar{x}+\delta x, u^*+\delta u, \tilde{V}_x(\bar{x}+\delta x, \omega; t); t) + \tilde{V}_x{}^T(\bar{x}+\delta x, \omega; t)w \\ &\quad - \tilde{H}(\bar{x}, \bar{u}, \tilde{V}_x(\bar{x}, \omega; t); t) - \tilde{V}_x{}^T(\bar{x}, \omega; t)w \\ &= \Delta\tilde{H}(t) + \delta x^T[\tilde{H}_x + \tilde{V}_{xx}\Delta f + \tilde{V}_{xx}(f(\bar{x}, \bar{u}; t)+w)] \\ &\quad + \delta u^T \tilde{H}_u + \tfrac{1}{2}\delta u^T \tilde{H}_{uu}\delta u + \delta u^T[\tilde{H}_{uu} + f_u{}^T \tilde{V}_{xx}]\delta x \\ &\quad + \tfrac{1}{2}\delta x^T[\tilde{H}_{xx} + f_x{}^T \tilde{V}_{xx} + \tilde{V}_{xx} f_x + \sum_{i=1}^{n} \tilde{V}^i_{xxx}\Delta f_i \\ &\quad + \sum_{i=1}^{n} \tilde{V}^i_{xxx}(f_i(\bar{x}, \bar{u}; t) + w_i)]\delta x + \cdots \end{aligned} \tag{6.3.20}$$

The unspecified arguments are $\bar{x}$, u^*, and $\tilde{V}_x(\bar{x}, \omega; t)$. Here, $\tilde{V}^i_{xxx}$ is a $n \times n$ matrix whose rth component is $(\partial^3/\partial x_i \partial x_r \partial x_s)[\tilde{V}(\bar{x}, \omega; t)]$ and

$$\Delta f = f(\bar{x}, u^*; t) - f(\bar{x}, \bar{u}; t) \tag{6.3.21}$$

The optimal δu minimizes the expectation, over all realizations of $w(\tau)$, $t \leqslant \tau \leqslant t_f$, of the right-hand side of Equation (6.3.21). Hence the optimal local control law is given by

$$\delta u(t) = \beta(t)\,\delta x(t) \tag{6.3.22}$$

$$\beta(t) = -H_{uu}^{-1}(\bar{x}; u^*; t)\,B(\bar{x}; u^*; t) \tag{6.3.23}$$

where

$$\begin{aligned} H_{uu}(\bar{x}, u^*; t) &= \underset{|x(t)=\bar{x}}{E} \tilde{H}_{uu}(\bar{x}, u^*, \tilde{V}_x; t) \\ &= L_{uu}(\bar{x}, u^*; t) + [V_x{}^T(\bar{x}; t) \cdot f(\bar{x}, u^*; t)]_{uu} \end{aligned} \tag{6.3.24}$$

($[\ \cdot\ ]_{uu}$ denote $(\partial/\partial u^2)[\ \cdot\ ]$):

$$\begin{aligned} B(\bar{x}, u^*; t) &= \underset{|x(t)=\bar{x}}{E} [\tilde{H}_{ux}(\bar{x}, u^*, V_x; t) + f_u{}^T(\bar{x}, u^*; t)\,\tilde{V}_{xx}] \\ &= L_{ux}(\bar{x}, u^*; t) + [V_x{}^T(\bar{x}; t) f(\bar{x}, u^*; t)]_{ux} \\ &\quad + f_u{}^T(\bar{x}, u^*; t)\,V_{xx}(\bar{x}; t) \end{aligned} \tag{6.3.25}$$

With the control law

$$u(t) = u^*(t) + \beta(t)[x(t) - \bar{x}(t)] \tag{6.3.26}$$

Equation (6.3.8) becomes, at $\tau = t$:

$$\begin{aligned} -\dot{\tilde{a}}(t) &= \Delta\tilde{H} \\ -\dot{\tilde{V}}_x(t) &= \tilde{H}_x + \beta^T \tilde{H}_u + \tilde{V}_{xx}\Delta f \\ -\dot{\tilde{V}}_{xx}(t) &= \tilde{A} + \beta^T \tilde{H}_{uu}\beta + \beta^T \tilde{B} + \tilde{B}\beta \end{aligned} \tag{6.3.27}$$

where

$$\tilde{A} = \tilde{H}_{xx} + f_x{}^T \tilde{V}_{xx} + \tilde{V}_{xx} f_x \tag{6.3.28}$$

The unspecified arguments are $\bar{x}$, u^*, $\tilde{V}_x(t)$, and t.

The naive Monte-Carlo algorithm consists of determining J realizations of $\tilde{V}_x(\bar{x}, \omega; t)$, $\tilde{V}_{xx}(\bar{x}; \omega; t)$, using Equations (6.3.1) and (6.3.4) with initial conditions $x(t) = \bar{x}$, $u(t) = \bar{u}$, and Equations (6.3.8) with final conditions given by Equations (6.3.9). The sample averages $\tilde{V}_x(\bar{x}; t)$, $\tilde{V}_{xx}(\bar{x}; t)$ are then determined; and, hence, using Equations (6.3.20), (6.3.26), and (6.3.27), $\hat{u}^*(t)$, $\hat{H}_{uu}(\bar{x}, u^*; t)$, $\hat{B}(\bar{x}, u^*; t)$, replacing expectations by the sample averages.

Then,

$$\hat{\beta}(t) = \hat{H}_{uu}^{-1}\hat{B} \tag{6.3.29}$$

The estimates u^* and β have random errors of $0(1/\sqrt{J})$. The standard deviation of the errors may bc estimated using the estimated variance of the various estimates.

Unlike the discrete-time algorithm, no "hill climbing," in the sense of generating new sets of realizations of $\bar{x}(\omega; t)$, is needed to determine u^*. All that is required is one set of J realizations, J sufficiently large to give the required degree of precision.

The first order algorithm uses only the first two equations of Equation (6.3.8) and yields an estimate of u^* only.

6.3.2. Determination of Optimal Parameters

The procedure is the same as that in Section 6.2.2.

$$\dot{\theta} = 0, \qquad \theta(t_o) = \bar{\theta} \tag{6.3.30}$$

replaces Equation (6.2.21), and $\tilde{V}_x(\bar{x}, \omega; t_o)$ and $\tilde{V}_{xx}(\bar{x}, \omega; t_o)$, corresponding to the realization $w(\omega; \tau)$, $t_o \leqslant \tau \leqslant t_f$ of the disturbance, are obtained using Equations (6.3.8) and (6.3.9); $\bar{x}$ may also be random.

6.3.3. Determination of Optimal (Open-Loop) Control

The problem is the obvious analog of the discrete-time problem discussed in Section 6.2.3. Let $\bar{U}$ specify the nominal control function $u(t)$, $t_o \leqslant t \leqslant t_f$, and U the new control function $u^*(t)$, $t_o \leqslant t \leqslant t_f$. Let $\bar{x}(\omega; t)$, $t_o \leqslant t \leqslant t_f$, denote the solution of the system equations with initial condition $x(\omega; t_o)$, control $\bar{U}$, and disturbance $w(\omega; t)$, $t_o \leqslant t \leqslant t_f$. The change in cost of realization ω due to replacing $\bar{U}$ by U is estimated by $\tilde{a}(x(\omega; t_o); t_o)$, the solution at t_o of

$$\begin{aligned} -\dot{\tilde{a}}(t) &= \Delta\tilde{H} \\ -\dot{\tilde{V}}_x(t) &= \tilde{H}_x \end{aligned} \tag{6.3.31}$$

with final conditions

$$\begin{aligned} \tilde{a}(t_f) &= 0 \\ \tilde{V}_x(t_f) &= F_x(\bar{x}(\omega; t_f)) \end{aligned} \tag{6.3.32}$$

The unspecified arguments are $\bar{x}(\omega; t)$, $u^*(t)$, $\tilde{V}_x(t)$, t. The terms are defined in Section 6.3.1. Hence, the change of cost is

$$\begin{aligned} \phi(U; t_o) - \phi(\bar{U}; t_o) &= E\,\tilde{a}(x(\omega; t_o); t_o) + O(\varepsilon^2) \\ &= \int_{t_o}^{t_f} E(\Delta\tilde{H})\,dt + O(\varepsilon^2) \end{aligned} \tag{6.3.33}$$

where the unspecified arguments are $\bar{x}(\omega; t)$, $u^*(t)$, $\tilde{V}_x(t)$, and t; and ε is the length of the interval over which $u^*(t)$ differs from $\bar{u}(t)$.

Hence an "ideal" algorithm would be to determine $u^*(t)$ according to

$$\begin{aligned} u^*(t) &= u \quad \text{that minimizes} \quad E\tilde{H}(\bar{x}(\omega; t), u, \tilde{V}_x(\bar{x}(\omega; t), \omega; t)) \\ &= u \quad \text{that minimizes} \quad EH(\bar{x}(\omega; t), u, V_x(\bar{x}(\omega; t); t)) \qquad (6.3.34) \end{aligned}$$

for $t_0 \leqslant t \leqslant t_f$. This new control is then applied over the interval $[t_1, t_f]$, t_1 being increased until an acceptability criterion is satisfied. A practical Monte-Carlo procedure replaces the expectation in Equation (6.3.34) by a sample average. The whole procedure, of course, has to be repeated until the estimate of $E\,\tilde{a}(t_0)$ is sufficiently small.

6.4. VARIANCE REDUCTION TECHNIQUES

The algorithms of Sections 6.2 and 6.3 are direct and obvious extensions of the deterministic algorithms. If these naive techniques are applied to linear systems, where the cost is quadratic in x and u, and the disturbance Gaussian (the LQG problem) it will be found that the estimates of the various variables required will all have a finite variance (called the sampling variance). Since the optimal solution for the LQG problem can be easily calculated, the Monte-Carlo method will not be competitive. A useful Monte-Carlo technique would yield zero variance estimates when applied to the LQG problem; such a technique would probably yield estimates of low-sampling variance when applied to problems that are approximately LQG. In this section we describe two variance reduction techniques that have been found useful [4, 5, 7] in control problems. For a fuller discussion of basic Monte-Carlo techniques, see Hammersley and Handscomb [3].

6.4.1. The Antithetic Variate Method

In the antithetic variate method, in its simplest form, two negatively correlated unbiased estimates $\hat{\theta}(+)$ and $\hat{\theta}(-)$ of a parameter θ are combined to give an estimate θ of smaller variance. Thus, if

$$\hat{\theta} = \tfrac{1}{2}\hat{\theta}(+) + \tfrac{1}{2}\hat{\theta}(-)$$

then the variance of $\hat{\theta}$ is

$$\begin{aligned} \operatorname{var}\hat{\theta} &= \tfrac{1}{4}\operatorname{var}\hat{\theta}(+) + \tfrac{1}{4}\operatorname{var}\hat{\theta}(-) \\ &\quad + \tfrac{1}{2}\operatorname{cov}(\hat{\theta}(+), \hat{\theta}(-)) \end{aligned}$$

If the estimates $\hat{\theta}(+)$ and $\hat{\theta}(-)$ have the same variance σ, and if $\text{cov}(\hat{\theta}(+), \hat{\theta}(-))$ is negative, then the variance of $\hat{\theta}$ is less than $\sigma/2$, the value it would have if $\hat{\theta}(+)$ and $\hat{\theta}(-)$ were independent estimates.

Consider the following model:

$$y = a + g(\omega)$$

where g is a skew-symmetric function and ω is a random variable with a symmetric probability density $p(\omega)$:

$$p(\omega) = p(-\omega)$$

The value of $\theta = Ey$ is desired. In naive Monte-Carlo calculations, J values of ω, drawn from $p(\omega)$, would be obtained, yielding J values of y:

$$y_j = a + g(\omega_j), \qquad j = 1, \ldots, J$$

The estimate is

$$\hat{\theta} = (1/J) \sum_{j=1}^{J} y_j$$

and has a variance σ:

$$\sigma = (1/J) \operatorname{var} g(\omega)$$

Using the antithetic variate technique, pairs of antithetic estimates are combined in the following way. Suppose ω_j^+ is a realization of ω, then, since $p(\omega)$ is symmetric, ω_j^- where

$$\omega_j^- = -\omega_j^+$$

is also a realization of ω with the same probability density. Hence, if y_j^+ and y_j^- are the corresponding values of y, then an unbiased estimate of θ is

$$\hat{\theta} = (1/J) \sum_{j=1}^{J} \tfrac{1}{2}(y_j^+ + y_j^-)$$

For our particular example, this reduces to

$$\begin{aligned} \hat{\theta} &= (1/J) \sum_{j=1}^{J} \tfrac{1}{2} (a + g(\omega_j^+) + a + g(\omega_j^-)) \\ &= (1/J) \sum_{j=1}^{J} \tfrac{1}{2}(a + g(\omega_j^+) + a - g(\omega_j^+)) \\ &= a \end{aligned}$$

since $g(\cdot)$ is skew symmetric. This estimate clearly has zero variance. In general, of course, $g(\cdot)$ is not skew symmetric, but the antithetic variate method removes all the sampling variance due to the skew-symmetric component of $g(\cdot)$. If $g(\omega)$ is linear in ω, the antithetic variate method yields zero sampling variance; we shall exploit this property in Section 6.5.

6.4.2. Control Variate Method

The control variate method of variance reduction can best be introduced by means of a simple example. Suppose that we wish to evaluate $Eg(\omega)$, where ω is a random variable of known distribution $\mathcal{F}(\omega)$. Using naive Monte-Carlo, a series of realizations ω_j, $j = 1, \ldots, J$, of ω is drawn from $\mathcal{F}(\omega)$. Then if

$$y = g(\omega) \tag{6.4.1}$$

the naive estimate $\hat{\theta}$ of $\theta = Eg(\omega)$ is

$$\hat{\theta} = (1/J) \sum_{j=1}^{J} y_j$$

and has a variance $(1/J)\operatorname{var} y$. In the control variate method a "model" (or approximation) of the observation equation (6.4.1) is formed, such that the expectation of the model output can be determined. For example, if the model is

$$\bar{y} = a + b\omega \tag{6.4.2}$$

then

$$E\bar{y} = a \tag{6.4.3}$$

if $\mathcal{F}(\omega)$ has zero mean. Now,

$$\begin{aligned} Ey &= E(y-\bar{y}) + E\bar{y} \\ &= E(y-\bar{y}) + a \end{aligned} \tag{6.4.4}$$

Naive Monte-Carlo is used to estimate $E(y-\bar{y})$, so that if $Ey = \theta$, then the control variate estimate is

$$\hat{\theta} = (1/J) \sum_{j=1}^{J} (y_j - \bar{y}_j) + a$$

Clearly the estimate is unbiased, and has a variance

$$\operatorname{var}\hat{\theta} = (1/J)\operatorname{var}(y-\bar{y}).$$

If $\bar{y}$ is a good approximation to y, then the variance of $(y-\bar{y})$ can be substantially lower than the variance of y; in such a case the control variate estimate of θ is substantially better than the naive estimate.

6.4.3. Conclusion

A variety of variance reduction techniques is discussed in the excellent book by Hammersley and Handscomb [3]. Because of the relative complexity of stochastic control problems, it seems that the above two methods, which may be used in combination, are of most use for this application.

6.5. ACCELERATED MONTE-CARLO GRADIENT METHODS FOR OPTIMIZING STOCHASTIC SYSTEMS

6.5.1. Introduction

In this section we describe briefly the application of the variance reduction techniques of Section 6.4 to the control algorithms of Sections 6.2 and 6.3 [4, 5]. The various parameters we require to estimate are:

1. (From Section 6.2.1):

$$\underset{|\bar{x}_k}{E}\, \tilde{H}_u^{\,k}, \qquad \underset{|\bar{x}_k}{E}\, \tilde{C}_k, \qquad \underset{|\bar{x}_k}{E}\, \tilde{B}_k$$

2. (From Section 6.2.2):

$$\underset{|\bar{x}_0}{E}\, \tilde{V}_x^{\,0}, \qquad \underset{|\bar{x}_0}{E}\, \tilde{V}_{xx}^{0}$$

or

$$E\tilde{V}_x^{\,0}, \qquad E\tilde{V}_{xx}^{0}$$

3. (From Section 6.2.3):

$$E\tilde{H}_u^{\,k}$$

4. (From Section 6.3.1):

$$\underset{|x(t)=\bar{x}}{E}\, \tilde{H}(t), \qquad \Big(\underset{|x(t)=\bar{x}}{E}\, \tilde{V}_x(t)\Big)$$

$$\underset{|x(t)=\bar{x}}{E}\, \tilde{H}_{uu}(t), \qquad \underset{|x(t)=\bar{x}}{E}\, \tilde{B}(t), \qquad \Big(\underset{|x(t)=\bar{x}}{E}\, \tilde{V}_{xx}(t)\Big)$$

5. (From Section 6.3.2):

$$\underset{|x(t_0)=\bar{x}}{E}\, \tilde{V}_x(t), \qquad \underset{|x(t_0)=\bar{x}}{E}\, \tilde{V}_{xx}(t)$$

or

$$E\tilde{V}_x(t), \qquad E\tilde{V}_{xx}(t)$$

6. (From Section 6.3.3):

$$E\tilde{H}(t), \qquad E\tilde{H}_u(t)$$

We shall also show in this section that the application of the variance reduction techniques to the LQG problem results, in most cases, in estimates of zero variance.

6.5.2. Estimation of Control Parameters Using the Antithetic Variate Method

The various algorithms are almost identical to the algorithms described in Sections 6.2 and 6.3. In each of these algorithms a random sequence S_j is generated, using a random number generator. This sequence specifies the initial condition (for the cases where the initial condition is random) and the disturbance sequence or function. Using this realization $(\omega = \omega_j)$ of the random sequence, an estimate $\theta(\omega_j)$ of the parameter ϕ is obtained. With J realizations of the random sequence, an improved estimate:

$$\hat{\phi} = (1/J) \sum_{j=1}^{J} \theta(\omega_j) \tag{6.5.1}$$

is obtained; ϕ denotes any of the quantities listed in Section 6.5.1.

To employ the antithetic variate method, we impose the condition that the probability density of each of the independent random variables in the sequence S is symmetric. Realizations of the sequence are now generated in pairs. To each realization S_j^+ of S, we generate an antithetic realization S_j^- as follows: If ξ is a member of the sequence S_j^+, the corresponding member of the sequence S_j^- is $-\xi$. The estimates of ϕ obtained from these two realizations are denoted by $\theta(\omega_j^+)$ and $\theta(\omega_j^-)$, respectively. With J pairs of realizations, the estimate $\hat{\phi}$ of ϕ is

$$\hat{\phi} = (1/J) \sum_{j=1}^{J} \tfrac{1}{2}[\theta(\omega_j^+) + \theta(\omega_j^-)] \tag{6.5.2}$$

Thus only a simple modification of the algorithms described in Sections 6.2 and 6.3 is required to implement the antithetic variate method. In each case a single realization S_j of the appropriate random sequence S is replaced by an antithetic pair, S_j^+ and S_j^-, constructed as described above, and the corresponding estimates of the parameter ϕ combined as shown in Equation (6.5.2).

6.5.3. Application of the Antithetic Variate Method to the LQS Problem

Consider the following system:

$$x_{i+1} = A_i x_i + B_i u_i + c_i + w_i \tag{6.5.3}$$

$$\begin{aligned} L_i &= \tfrac{1}{2} x_i^T Q_i x_i + q_i^T x_i \\ &+ \tfrac{1}{2} u_i^T R_i u_i + r_i^T u_i + u_i^T S_i x_i \end{aligned} \tag{6.5.4}$$

$$F = \tfrac{1}{2} x_N^T Q_N x_N + q_N^T x_N \tag{6.5.5}$$

where $\{w_0, ..., w_{N-1}\}$ is a sequence of independent random variables of zero mean and symmetric probability density. Also, the initial condition is either deterministic or a random variable with symmetric probability density.

Assume that the initial condition (deterministic or random) is:

$$x_k = \bar{\bar{x}}_k + \xi_k \tag{6.5.6}$$

where k is possibly zero, and either $p(\xi_k)$ is symmetric or ξ_k is zero if the initial condition is deterministic; and that the control policy for $i > k$ is linear:

$$u_i = x_i + \beta_i x_i, \qquad i = k+1, ..., N-1 \tag{6.5.7}$$

We shall refer to a system having all these properties as an LQS system. Clearly,

$$\bar{x}_i(\omega) = \bar{\bar{x}}_i + \xi_i, \qquad i = k+1, ..., N \tag{6.5.8}$$

where k is possibly zero, and ξ_i is a random variable of zero mean and symmetric probability density; $\bar{\bar{x}}_j$, $j = k+1, ..., N$, is the solution of Equations (6.5.3) and (6.5.7) with initial condition $\bar{\bar{x}}_k$ and $w_j = 0, j = k, ..., N-1$. The difference equations for $\tilde{V}_x^{\,i}$, $\tilde{V}_{xx}^{i}$ become (see Equations (6.2.6) and (6.2.7)):

$$\begin{aligned} \tilde{V}_x^{\,i} &= Q_i \bar{\bar{x}}_i + Q_i \xi_i + q_i \\ &\quad + [A_i + B_i \beta_i]^T \tilde{V}_x^{i+1} \end{aligned} \tag{6.5.9}$$

$$\tilde{V}_{xx}^{i} = Q_i + [A_i + B_i \beta_i]^T \tilde{V}_{xx}^{i+1} [A_i + B_i \beta_i] \tag{6.5.10}$$

with terminal conditions:

$$\tilde{V}_x^{\,N} = Q_N \bar{\bar{x}}_N + Q_N \xi_N + q_N \tag{6.5.11}$$

$$\tilde{V}_{xx}^{N} = Q_N \tag{6.5.12}$$

Now

$$\xi_i = \sum_{j=k}^{i-1} G_j w_j + H \xi_k, \qquad i = k+1, ..., N \tag{6.5.13}$$

where the matrices G_j and H depend on A_j, B_j, c_j, and β_j, $j = k+1, ..., N-1$. Hence the antithetic realizations of ξ_i are related by

$$\xi_i(\omega_j^{\,+}) = -\xi_i(\omega_j^{\,-}), \qquad i = k+1, ..., N \tag{6.5.14}$$

Hence, from Equations (6.5.9) and (6.5.11), the values of $\tilde{V}_x^i$, obtained using the jth pair of antithetic realizations, are related as follows:

$$[\tilde{V}_x^i(\omega_j^-) - E\tilde{V}_x^i] = -[\tilde{V}_x^i(\omega_j^+) - E\tilde{V}_x^i], \qquad i = k, ..., N \quad (6.5.15)$$

Hence, since $\tilde{H}_u^i = R_i u_i + r_i + S_i x_i + B_i^T \tilde{V}_x^{i-1}$, we have

$$[\tilde{H}_u^i(\omega_j^-) - EH_u^i] = -[\tilde{H}_u^i(\omega_j^+) - E\tilde{H}_u^i], \qquad i = k, ..., N \quad (6.5.16)$$

We also note that Equations (6.5.10) and (6.5.12) do not depend on the random variables ξ_k, $w_k, ..., w_{N-1}$. Hence, application of the antithetic variate Monte-Carlo technique to an LQS discrete-time system yields zero variance estimates of the following quantities:

$$\underset{|x_k=\bar{x}_k}{E}\tilde{V}_x^k, \qquad \underset{|x_k=\bar{x}_k}{E}\tilde{V}_{xx}^k, \qquad \underset{|x_k=\bar{x}_k}{E}\tilde{H}_u^k, \qquad \underset{|x_k=\bar{x}_k}{E}\tilde{A}_k$$

$$\underset{|x_k=\bar{x}_k}{E}\tilde{B}_k, \qquad \underset{|x_k=\bar{x}_k}{E}\tilde{C}_k$$

$$E\tilde{V}_x^k, \qquad E\tilde{V}_{xx}^k, \qquad E\tilde{H}_u^k, \qquad E\tilde{A}_k$$

$$E\tilde{B}_k, \qquad E\tilde{C}_k, \qquad k = 0, ..., N-1$$

Hence, application of this technique to an LQS system yields zero variance estimates of the control parameters required in Sections 6.2.1 to 6.2.3. Note that, for an LQS system, application of a naive Monte-Carlo technique yields zero variance estimates of the second-order quantities,

$$\underset{|x_k=\bar{x}_k}{E}\tilde{V}_{xx}^k, \; E\tilde{V}_{xx}^k, \text{ etc.}$$

Consider now the continuous-time LQS system:

$$\dot{x}(\tau) = A(\tau)x(\tau) + B(\tau)u(\tau) + c(\tau) + w(\tau) \quad (6.5.17)$$

$$L(x, u; \tau) = \tfrac{1}{2}x^T Q(\tau)x + q^T(\tau)x + \tfrac{1}{2}u^T R(\tau)u + r^T(\tau)u + u^T S(\tau)x \quad (6.5.18)$$

$$F(x) = \tfrac{1}{2}x^T Q_f x + q_f^T x \quad (6.5.19)$$

$$u(\tau) = \alpha(\tau) + \beta(\tau)x(\tau), \qquad t < \tau \leqslant t_f \quad (6.5.20)$$

where the initial condition is

$$x(t) = \bar{\bar{x}}(t) + \xi(t) \quad (6.5.21)$$

t is possibly equal to t_0, and $\xi(t)$ is zero if the initial condition is deterministic. It is assumed that $\xi(t)$ and $w_i, w_{i+1}, \ldots$ [the random variables that specify $w(\tau)$] are independent random variables of zero mean and symmetric probability density. As before,

$$x(\tau) = \bar{\bar{x}}(\tau) + \xi(\tau), \qquad t \leqslant \tau \leqslant t_f \tag{6.5.22}$$

where $\xi(\tau)$ is given by

$$\xi(\tau) = H\xi(t) + G_i w_i + G_{i+1} w_{i+1} + \cdots \tag{6.5.23}$$

The differential equations for $\tilde{V}_x$ and $\tilde{V}_{xx}$ are

$$\begin{aligned} -\dot{\tilde{V}}_x(\tau) &= Q(\tau)\bar{\bar{x}}(\tau) + Q(\tau)\xi(\tau) + q(\tau) \\ &\quad + [A(\tau) + B(\tau)\beta(\tau)]^T \tilde{V}_x(\tau) \end{aligned} \tag{6.5.24}$$

$$-\dot{\tilde{V}}_{xx}(\tau) = Q(\tau) + [A(\tau) + B(\tau)\beta(\tau)]^T \tilde{V}_{xx}(\tau)[A(\tau) + B(\tau)\beta(\tau)] \tag{6.5.25}$$

with boundary conditions:

$$\tilde{V}_x(t_f) = Q_f \bar{\bar{x}}(t_f) + Q_f \xi(t_f) + q_f(t_f) \tag{6.5.26}$$

$$\tilde{V}_{xx}(t_f) = Q_f \tag{6.5.27}$$

It follows immediately that

$$\tilde{V}_x(\omega_j^-; t) - E\tilde{V}_x(t) = -[\tilde{V}_x(\omega_j^+; t) - E\tilde{V}_x(t)] \tag{6.5.28}$$

$$\tilde{V}_{xx}(\omega_j; t) = \tilde{V}_{xx}(\omega_k; t), \qquad \omega_j = \omega_j^+ \quad \text{or} \quad \omega_j^- , \quad \text{etc.} \tag{6.5.29}$$

for the LQS system. For this system,

$$\begin{aligned} \tilde{H} &= \tfrac{1}{2} x^T Q(t) x + q^T(t) x + \tfrac{1}{2} u^T R(t) u \\ &\quad + r^T(t) u + u^T S(t) x \\ &\quad + \tilde{V}_x{}^T(t) \cdot [A(t)x(t) + B(t)u(t) + c(t)] \end{aligned} \tag{6.5.30}$$

Hence, application of the antithetic variate Monte-Carlo technique to an LQS continuous-time system yields zero variance estimates of the following quantities:

$$\underset{|x(t)=\bar{x}}{E}\ \tilde{V}_x(t), \qquad \underset{|x(t)=\bar{x}}{E}\ \tilde{H}(t), \qquad \underset{|x(t)=\bar{x}}{E}\ \tilde{V}_{xx}(t)$$

$$\underset{|x(t)=\bar{x}}{E}\ \tilde{H}_u(t), \qquad \underset{|x(t)=\bar{x}}{E}\ \tilde{H}_{uu}(t)$$

$$\underset{|x(t)=\bar{x}}{E}\ \tilde{H}_{ux}(t), \qquad E\tilde{H}_u(t), \qquad E\tilde{V}_{xx}(t), \qquad E\tilde{V}_x(t)$$

$$E\tilde{H}_{uu}(t), \qquad E\tilde{H}_{ux}(t), \qquad E\tilde{H}_{xx}(t) \qquad \text{for} \quad t_0 \leqslant t \leqslant t_f$$

Consequently, application of this technique to a LQS system results in zero variance estimates of the control parameters required in Sections 6.3.1 to 6.3.3. [However, for determining an optimal open-loop control function, the minimum of $E\tilde{H}(t)$ is obtained using the necessary condition $E\tilde{H}_u(t) = 0$, since the method *does not* yield a zero variance estimate of $E\tilde{H}(t)$.]

6.5.4. Control Variate Technique

To employ the control variate technique of variance reduction, an model of the original process has to be obtained. Assume therefore that we have obtained a linear model of the nonlinear stochastic system. This model may be obtained, e.g., by linearization about the trajectory obtained using (the mean value of) the initial condition and setting the disturbances equal to zero. Alternatively, statistical linearization may be employed to yield a linear model; such a model has been used (Handschin and Mayne [6]) to reduce the sampling variance of Monte-Carlo procedures for nonlinear filtering. The resultant linear model may be used in the following way. Assume that an estimate of $E[\tilde{V}_x(t)|x(t) = \bar{x}]$ is required. Then, using the method already described, realizations $\tilde{V}_x(\omega_k; t), j = 1, \ldots, J$ are obtained. The *same* method is now applied to the linear model, using the same initial condition and the same disturbance (specified by ω), to yield ${}_m\tilde{V}_x(\omega_j; t)$, where m denotes a variable of the linear model. If the model is a good approximation to the linear system, ${}_m\tilde{V}_x(\omega_j; t)$ will be a good approximation to $\tilde{V}_x(\omega_j; t)$. Naive Monte-Carlo is now used to estimate the difference:

$$\underset{|x(t)=\bar{x}}{E}\ [\tilde{V}_x(t) - {}_m\tilde{V}_x(t)] \tag{6.5.31}$$

Since the model is linear, the expectation $E[{}_m\tilde{V}_x(t)|x(t) = \bar{x}]$ can be calculated, yielding finally the following Monte-Carlo estimate of $E[\tilde{V}_x(t)|$ $x(t) = \bar{x}]$:

$$\hat{V}_x(\bar{x}; t) = {}_mV_x(\bar{x}; t) + (1/J) \sum_{j=1}^{J} [\tilde{V}_x(\omega_j; t) - {}_m\tilde{V}_x(\omega_j; t)] \tag{6.5.32}$$

where

$$_{m}V_x(\bar{x};t) = \underset{|x(t)=\bar{x}}{E} \; _{m}\tilde{V}_x(t) \tag{6.5.33}$$

If $_{m}\tilde{V}_x(t)$ is a good approximation to $\tilde{V}_x(t)$, the difference term in Equation (6.5.32) is small, and the resultant variance of the estimate $\hat{V}_x(\bar{x};t)$ will be low. The same procedure can be used to estimate any other quantity.

It is obvious that application of this procedure to the LQS problem will result in a linear model identical to the original linear system. Hence, $_{m}\tilde{V}_x(\omega_j;t)$ will be exactly equal to $\tilde{V}_x(\omega_j;t)$, the resultant difference will be zero, and the variance of the estimate $\hat{V}_x(\bar{x};t)$ will be zero. Similarly, any other estimate using this technique will have zero variance.

6.6. OPTIMAL LINEAR CONTROL POLICY

An optimal, linear control policy can be obtained using the method described in Sections 6.2.3, 6.3.3, the control at time k (time t) specifying the parameters of a feedback control. If we consider the discrete-time control system with a scalar control, and the kth control law defined by

$$u_k = \alpha_k + \beta_k^T x_k \tag{6.6.1}$$

then the gradient of the cost function with respect to the vector θ_k;

$$\theta_k^T = (\alpha_k, \beta_k^T) \tag{6.6.2}$$

is

$$\begin{bmatrix} E\tilde{H}_u^k \\ E\tilde{H}_u^k x_k \end{bmatrix} \tag{6.6.3}$$

and the second derivative of the cost function is

$$\begin{bmatrix} E\tilde{C}_k & E\tilde{C}_k x_k^T \\ E\tilde{C}_k x_k & E\tilde{C}_k x_k x_k^T \end{bmatrix} \tag{6.6.4}$$

If the system is LQS, all these quantities can be estimated with zero variance using the control variate method. Using J realizations, $J > n$, the optimal, linear control policy of an LQS system can be obtained by one "backward" integration of the J realizations of the $\tilde{V}_x^k$ and $\tilde{V}_{xx}^k$ sequences. Of course, the difference equations for the $\tilde{V}_x^k$ and $\tilde{V}_{xx}^k$ sequences uses the values of α_k and β_k determined during the backward integration. For further details, see Westcott *et al.* [5].

6.7. STOCHASTIC SYSTEMS WITH CONTROL CONSTRAINTS

Consider the problem of determining the optimal open-loop control of the continuous-time system discussed in Section 6.3, where now the control is subject to constraints of the form:

$$u^a \leqslant u \leqslant u^b \tag{6.7.1}$$

Let the set satisfying Equation (6.7.1) be denoted by U. Then it is easily shown that if $\bar{u}(t)$, $t_o \leqslant t \leqslant t_f$, is a nominal control, resulting in the nominal trajectory $\bar{x}(\omega; t)$, $t_o \leqslant t \leqslant t_f$, for each $\omega \in \Omega$, an improved control is

$$u(t) = \bar{u}(t), \qquad t_o \leqslant t \leqslant t_\alpha$$

$$u(t) = u^*(t), \qquad t_\alpha \leqslant t \leqslant t_f$$

where $(t_f - t_\alpha)$ is sufficiently small and where $u^*(t)$ is that value of $u \in U$ that minimizes

$$E\tilde{H}(\bar{x}(\omega; t), u, \tilde{V}_x(t); t) \tag{6.7.2}$$

where $\tilde{a}(t)$ and $\tilde{V}_x(t)$ are the solutions at time t of

$$-\dot{\tilde{a}}(t) = \Delta\tilde{H}(t) \tag{6.7.3}$$

$$-\dot{\tilde{V}}_x(t) = \tilde{H}_x(t) \tag{6.7.4}$$

$$\tilde{a}(t_f) = 0 \tag{6.7.5}$$

$$\tilde{V}_x(t_f) = F_x(\bar{x}(\omega; t_f)) \tag{6.7.6}$$

the unspecified arguments being $\bar{x}(\omega; t)$, u^*, $V_x(t)$, and t.

If this algorithm is applied to an LQS system, then the antithetic variate method results in a zero sampling variance estimate of:

$$E\tilde{H}_u(t) = R(t)u(t) + r(t) + B^T(t)\tilde{V}_x(t) + S(t)\bar{x}(t)$$

but not of $E\tilde{H}(t)$. However, the gradient $E\tilde{H}_u(t)$ can be used to determine $u^*(t)$. Hence the antithetic variate method, if applied to the LQS system, results in an estimate of a (locally) optimal open-loop control with zero sampling variance

6.8. SOME EXAMPLES

1. Consider the nonlinear problem defined by†:

$$x_{k+1} = x_k + f(u_k) + w_k, \qquad L_k = \tfrac{1}{2}(x_k^2 + u_k^2)$$

$$F = \tfrac{1}{2} x_N^2$$

where $f(\cdot)$ is defined by

$$f(u) = u, \qquad |u| \leqslant 1; \qquad f(u) = 2 - \exp(1-u), \qquad u > 1$$

$$f(u) = -2 + \exp(1+u), \qquad u < -1$$

$\{w_k\}$ is a sequence of independent Gaussian random variables of zero mean and unit variance, $N = 10$. Using the antithetic variate technique, the variance of $E\tilde{H}_u^k$ was reduced by a factor of 50 to 5000 compared with naive Monte-Carlo for those values of k where the system was operating in the nonlinear region. In the linear region the reduction factor, theoretically infinite, was found to be from 10^{11} to 10^{14}. It was found, using the antithetic variate method, that $J = 10$ was sufficient to obtain sufficiently precise estimates of the optimal open-loop control.

2. The second-order method of Section 6.2.1 was then applied‡ to the above problem, with $N = 10$, to determine the optimal control action for $0 \leqslant x_9 \leqslant 10$. Using the antithetic variate method $J = 10$ was sufficient to determine the optimal control law. The result was a smooth curve differing very little from the "curve" abcd in Figure 6.1. The second-order procedure of Section 6.6 was then used to get a sequence of optimal linear control laws. For $x_0 = 14$, 18, and 22, the optimal linear control law at $k = 9$ is shown in Figure 6.1 by, respectively, the lines af, ch, gd. The projection of these lines onto the horizontal axis indicates the range of the realization of x_{10} for the three initial conditions. The antithetic variate technique was used.

3. To illustrate the application of the control variate method the following problem, which has a quadratic nonlinearity, was studied:

$$x_{k+1} = x_k + 0.125\, x_k^2 + u_k + 0.1\, w_k$$

L_k, F and $\mathrm{var}(w_k)$ are the same as in problem (1). Also,

$$x_0 = 0, \qquad N = 10$$

† See Mayne [4].

‡ See Westcott *et al.* [5].

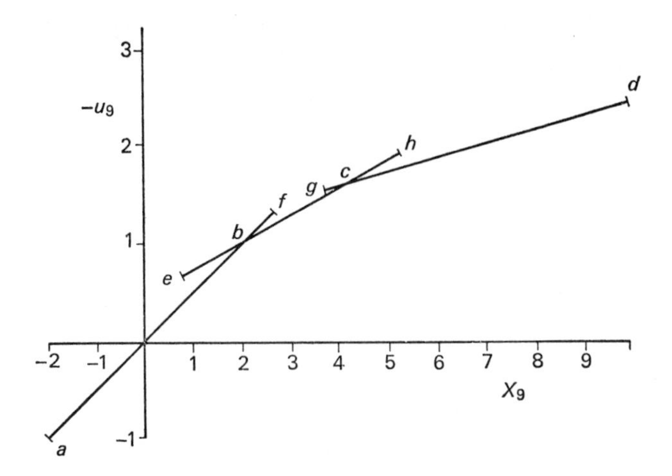

Figure 6.1

The optimal deterministic solution ($w_k = 0$, $k = 0, \ldots, 9$) was determined ($x_k = 0$, $k = 0, \ldots, 10$; $u_k = 0$, $k = 0, \ldots, 9$) and the system linearized about this solution to yield a linear model. The optimal policy for the linear model was then determined:

k	0	1	2	3	4	5	6	7	8	9
α_k	0	0	0	0	0	0	0	0	0	0
β_k	−0.618	−0.618	−0.618	−0.618	−0.618	−0.618	−0.618	−0.615	−0.6	−0.5

These values were used as the nominal values of a linear control policy. The linear model was then used to predict the variance quantities required to improve this linear policy. Some of the results from one realization are shown in Table I.

Table I

k	0	1	2	3	4	5	6	7	8	9	10
$\lvert x_k \rvert$	0	0.0103	0.0514	0.1357	0.0289	0.1150	0.0206	0.1368	0.0267	0.0072	0.0216
$\lvert {}_m x_k \rvert$	0	0.0103	0.0532	0.1364	0.0275	0.1150	0.0209	0.1343	0.0271	0.0057	0.0203
$\lvert \tilde{V}_x^k \rvert$	0.0259	0.0678	0.1393	0.1729	0.0352	0.1928	0.0954	0.1787	0.0282	0.0197	0.0216
$\lvert {}_m\tilde{V}_x^k \rvert$	0.0257	0.0673	0.1390	0.1745	0.0366	0.1953	0.0953	0.1736	0.0300	0.0173	0.0203
$\lvert \tilde{H}_u^k x_k \rvert$	0	0.0014	0.0073	0.0161	0.0061	0.0029	0.0034	0.0151	0.0009	0.0001	
$\lvert {}_m\tilde{H}_u^k \cdot {}_m x_k \rvert$	0	0.0014	0.0074	0.0165	0.0058	0.0028	0.0034	0.0151	0.0009	0.0000	
$100\lvert \tilde{C}_k(x_k)^2 \rvert$	0	0.0284	0.6732	4.653	0.1978	3.451	0.1296	5.105	0.1362	0.0193	
$100\lvert {}_m\tilde{C}_k({}_m x_k)^2 \rvert$	0	0.0277	0.6911	4.795	0.1835	3.523	0.1345	4.924	0.1424	0.0139	

The close correspondence between the actual and predicted quantities indicates the variance reduction obtained. As an example, for $k = 1$, the following quantity was calculated using linear system theory:

$$E_m \tilde{C}_1 ({}_m x_1)^2 = 0.02618$$

The control variate estimate of $E\, \tilde{C}_1\, x_1{}^2$ is

$$0.02618 + 0.00001$$

where only the smaller term had to be evaluated using Monto-Carlo techniques. It was found that, for $J = 10$, the variance of the estimate was 1.275×10^{-10}, corresponding to a standard deviation of 0.00001 that is satisfactorily small compared with 0.02619, (though large compared with 0.00001, the estimated smaller term).

6.9. CONCLUSION

The examples, though trivial, indicate the possible usefulness of Monte-Carlo techniques for optimal control problems. Monte-Carlo techniques seem particularly suited to those problems difficult to solve by other methods, namely, the determination of optimal open-loop control or the determination of optimal parameters of a parameterized control law. The antithetic variate method of variance reduction is very simple to apply, and often yields substantial improvement in the accuracy of the estimates. The control variate method requires a linearized model of the nonlinear system. This model is often obtained anyway to yield an approximate solution to the control problem. The control variate method can be regarded as a Monte-Carlo procedure to improve this approximate solution.

The procedures given in Chapters 5 and 6 have sidestepped some difficult theoretical problems. The algorithms, in effect, optimize a model of the original stochastic system, the model being capable of simulation on a digital computer. The construction of such models, the closeness of their approximation to the original system and, more important, the closeness of the resultant optimal control of the model to the optimal control of the original system, all require further study. Another topic not covered is the development of techniques for systems with partially accessible states.

References

1. J. M. C. Clark, Ph. D. Thesis, Univ. of London, England, 1966.
2. E. Wong and M. Zakai, *Intern. J. Eng. Soc.* **3**, 213 (1965).
3. D. M. Hammersley and D. C. Handscomb, "Monte Carlo Methods, "Methuen, London, 1965.

4. D. Q. Mayne, *in Proc. 2nd IFAC Symp. on Theory of Self-Adaptive Control Systems* Plenum Press, New York, 1966.
5. J. H. Westcott, D. Q. Mayne, G. F. Bryant, and S. K. Mitter, *in Proc. 3rd IFAC Congress, London 1966.*
6. E. Handschin and D. Q. Mayne, *Intern. J. Control*, 9, 547, (1969).
7. D. H. Jacobson and D. Q. Mayne, *Proc. 4th Congr. of Intern. Fed. Auto. Control, Warsaw, 1969.*

APPENDIX

Implicit in all the derivations of this chapter is the assumption that the order of differentiation and expectation can be interchanged. We assume, e.g., that

$$V_x(x; t) = E\tilde{V}_x(x, \omega; t)$$

i.e.,

$$(\partial/\partial x)[E\tilde{V}(x, \omega; t)] = E[(\partial/\partial x)\tilde{V}(x, \omega; t)]$$

Since we are, in effect, determining the optimal control of a digital computer model of the original system, and since the number of possible realizations of the simulated random variables, though large, is still finite, the assumption is justified. However, we are requiring one further feature of our model; the properties of the original system should be adequately simulated using the discrete distributions of the simulated random variables. Nevertheless, the order of differentiation and expectation may be changed under fairly general conditions as shown by the following result (for which the authors are indebted to J.M.C. Clark): assume:

$$E\tilde{V}(a, \omega) < \infty \tag{6.A.1}$$

$$E\int_a^b |\tilde{V}_x(x, \omega)|\, dx < \infty \tag{6.A.2}$$

Then,

$$(\partial/\partial x)[E\tilde{V}(x, \omega)] = E[(\partial/\partial x)\tilde{V}(x, \omega)]$$

almost everywhere on $[a, b]$.

From (6.A.2), $\tilde{V}_x(x, \omega) \in L^1$ for almost all $\omega \in \Omega$ (6.A.3)

Also, from (6.A.2)and Fubini's theorem,

$$\int_a^x [E\tilde{V}_x(x', \omega)]\, dx' = E[\int_a^x \tilde{V}_x(x', \omega)\, dx'] \tag{6.A.4}$$

From (6.A.3) and the fundamental theorem of calculus,

$$\int_a^x \tilde{V}_x(x', \omega)\, dx' = \tilde{V}(x, \omega) - \tilde{V}(a, \omega) \tag{6.A.5.}$$

for almost all $\omega \in \Omega$. Hence,

$$\int_a^x [E\tilde{V}_x(x', \omega)]\, dx' = E\tilde{V}(x, \omega) - E\tilde{V}(a, \omega) \tag{6.A.6}$$

From (6.A.1), (6.A.2), and (6.A.4),

$$E\tilde{V}(x, \omega) < \infty \tag{6.A.7}$$

But, for every integrable function f,

$$(\partial/\partial x) \int_a^x f(x')\, dx' = f(x) \tag{6.A.8}$$

almost everywhere on $[a, b]$.
Identifying $E\tilde{V}_x(x, \omega)$ with $f(x)$, we have from (6.A.8),

$$E\tilde{V}_x(x, \omega) = (\partial/\partial x)[E\tilde{V}(x, \omega)]$$

almost everywhere on $[a, b]$.

Chapter 7

CONCLUSION

7.1. THE SIGNIFICANCE OF DIFFERENTIAL DYNAMIC PROGRAMMING

Differential dynamic programming is a useful, new computational technique for solving nonlinear dynamic optimization problems. By applying the principle of optimality locally, in the neighborhood of a nominal trajectory, we have been able to derive successive approximation algorithms that converge rapidly, for a large class of problems, to a local minimizing solution. Storage requirements of these algorithms are modest when compared with conventional dynamic programming. What we have sacrificed (relative to conventional dynamic programming) is, of course, global optimality and ease of dealing with state variable inequality constraints.

Restricting one's attention to the immediate neighborhood of a nominal trajectory is a common device and one used in the calculus of variations. However, algorithms as general and as powerful as those described in this book have not been forthcoming from variational analyses. This is attributed to the fact that the variational techniques do not, inherently, possess the great intuitive appeal that dynamic programming offers, and which was instrumental in the development of the novel differential dynamic programming algorithms.

The closest "relative" of differential dynamic programming appears to be the "successive-sweep" (or second-variation) method. We have shown, however, that this method in its current form is not applicable to as wide a class of problems as the algorithms described in Chapters 2 and 3, nor is it equivalent to the simplest differential dynamic programming algorithm (Appendix A). There is little doubt, however, that with hindsight, variational methods will be used to produce algorithms equivalent to those derived in this book.

A significant advance of differential dynamic programming is that second-order algorithms have been derived that are able to solve problems with control variable inequality constraints (in particular, bang-bang control problems). Implicit in the derivation and statement of the second-order algorithms for bang-bang control problems are sufficient conditions of optimality. It is believed that sufficient conditions of optimality for this class of problems have not been obtained heretofore. The derivation of these conditions, as given in this book, is essentially formal; however, it should be possible to provide rigorous proofs.

The relative simplicity of the differential dynamic programming algorithms also makes them useful for stochastic problems. Indeed, for the determination of an optimal set of parameters, or an optimal open-loop control function, the deterministic algorithms allied with Monte-Carlo simulations of the stochastic system provide useful and powerful procedures. For this type of problem, conventional dynamic programming is not very suitable. Experience with the deterministic algorithms, using global variations in control, has shown that rapid convergence can be obtained even if the first-order method is employed. The first-order algorithm for stochastic problems has modest storage requirements, since realizations do not have to be generated simultaneously.

7.2. EXTENSIONS OF DIFFERENTIAL DYNAMIC PROGRAMMING: DETERMINISTIC SYSTEMS

Extensions of differential dynamic programming to singular control problems and control problems with state variable inequality constraints are currently being investigated. Thus far, we have obtained new, necessary conditions of optimality for singular problems [1], and state constrained problems [2].

The theoretical and computational implications of higher-order expansions for V are subjects of current interest [3]. An important field yet to be studied in detail is the convergence properties of the various algorithms.

7.3. EXTENSIONS OF DIFFERENTIAL DYNAMIC PROGRAMMING: STOCHASTIC SYSTEMS

An obvious desirable extension is the development of methods for systems with noisy observations. Monte-Carlo methods for nonlinear filtering [4] may be of use, though it is possibly more fruitful to express the control directly as a function of the observations rather than as a function of the conditional probability distribution of the state. Problems associated with stochastic bang-bang systems and terminal constraints are also being investigated.

References

1. D. H. Jacobson, Harvard Univ. Tech. Rept, **TR 576** (November 1968).
2. D. H. Jacobson, M. M. Lele and J. L. Speyer, Havard Univ. Tech. Rept. **TR 591** (August 1969).
3. S. B. Gershwin, *J. Math. Anal. Appl.*, 28, 120 (1969).
4. E. Handschin and D. Q. Mayne, *Intern, J. Control*, 9, 547 (1969).

Appendix A

SECOND-ORDER AND SUCCESSIVE-SWEEP ALGORITHMS: A COMPARISON

A.1. INTRODUCTION

Mayne [2] has derived a second-order algorithm using the notion of differential dynamic programming. The algorithm is different from those described in Chapters 2 and 3 in that only small changes are made in the control function from iteration to iteration (i.e., weak variations in the x trajectory). Mitter [3] and McReynolds and Bryson [4] have obtained a similar, though nonequivalent, algorithm using the Lagrange multiplier technique. Recently, McReynolds [5] obtained a method equivalent to Mayne's.

Here we shall indicate some differences between the method described by Mayne [2] and that described by Mitter [3] and McReynolds and Bryson [4].‡ A more detailed comparison is found in Jacobson [1], where it is shown that the algorithm of Mitter, and McReynolds and Bryson, is an approximation to Mayne's.

A.2. A SECOND-ORDER WEAK VARIATION ALGORITHM

This algorithm, due to Mayne [2], is described below. The control problem is: choose $u(t)$; $t \in [t_o, t_f]$ to minimize

$$V(x_o; t_o) = \int_{t_o}^{t_f} L(x, u; t)\, dt + F(x(t_f)) \tag{A.2.1}$$

where the dynamic system is described by the differential equations:

$$\dot{x} = f(x, u; t); \qquad x(t_o) = x_o \tag{A.2.2}$$

The final time t_f is given explicitly.

‡ We shall consider the unconstrained, continuous-time control problem.

Using a differential dynamic programming approach, Mayne obtained the following equations:

$$-\dot{a} = -\varepsilon(1-\varepsilon/2)\langle H_u, H_{uu}^{-1} H_u\rangle \tag{A.2.3}$$

$$-\dot{V}_x = H_x + \beta^T H_u \tag{A.2.4}$$

$$-\dot{V}_{xx} = H_{xx} + f_x^T V_{xx} + V_{xx} f_x - (H_{ux} + f_u^T V_{xx})^T H_{uu}^{-1}(H_{ux} + f_u^T V_{xx}) \tag{A.2.5}$$

where

$$\beta = -H_{uu}^{-1}(H_{ux} + f_u^T V_{xx}) \tag{A.2.6}$$

and ε is a scalar, $0 < \varepsilon \leqslant 1$.

All the above quantities are evaluated *along the nominal trajectory* $\bar{x}(t)$, $\bar{u}(t)$; $t \in [t_o, t_f]$.

The new control at each iteration is given by

$$\begin{aligned} u(t) &= \bar{u}(t) - H_{uu}^{-1}[\varepsilon H_u + (H_{ux} + f_u^T V_{xx})\,\delta x] \\ &= \bar{u}(t) - \varepsilon H_{uu}^{-1} H_u + \beta\,\delta x \end{aligned} \tag{A.2.7}$$

For ε sufficiently small, and assuming $H_{uu}(\bar{x}, \bar{u}, V_x; t)$ is positive-definite, a reduction in cost is predicted at each iteration [i.e., $a(x_o; t_o) \leqslant 0$].

Boundary conditions for Equations (A.2.3) to (A.2.5) are

$$\begin{gathered} a(t_f) = 0 \\ V_x(t_f) = F_x(\bar{x}(t_f)), \qquad V_{xx}(t_f) = F_{xx}(\bar{x}(t_f)) \end{gathered} \tag{A.2.8}$$

A.3. THE SUCCESSIVE-SWEEP ALGORITHM

Mitter [3] and McReynolds and Bryson [4] obtained the algorithm described below by expanding the canonical equations to first-order in δx, $\delta\lambda$, and δu around a nominal trajectory; the resulting linear, two-point boundary-value problem is solved using the Riccati transformation (i.e., it is assumed that $\delta\lambda = h + P\,\delta x$).

$$-\dot{\lambda} = H_x \tag{A.3.1}$$

$$-\dot{h} = f_x^T h + \beta_1^T(H_u + f_u^T h) \tag{A.3.2}$$

$$-\dot{P} = H_{xx} + f_x^T P + P f_x - (H_{ux} + f_u^T P)^T H_{uu}^{-1}(H_{ux} + f_u^T P) \tag{A.3.3}$$

where H is defined by

$$H(x, u, \lambda; t) = L(x, u; t) + \langle \lambda, f(x, u; t)\rangle \tag{A.3.4}$$

and

$$\beta_1 = -H_{uu}^{-1}(H_{ux} + f_u{}^T P) \tag{A.3.5}$$

All quantities are evaluated *along the nominal trajectory* $\bar{x}(t)$, $\bar{u}(t)$; $t \in [t_o, t_f]$.
The control at each iteration is given by

$$u(t) = \bar{u}(t) - \varepsilon H_{uu}^{-1}(H_u + f_u{}^T h) + \beta_1 \delta x \tag{A.3.6}$$

Boundary conditions for Equations (A.3.3) to (A.3.5) are

$$\lambda(t_f) = F_x(\bar{x}(t_f)), \qquad h(t_f) = 0$$
$$P(t_f) = F_{xx}(\bar{x}(t_f)) \tag{A.3.7}$$

For ε sufficiently small $(0 \leqslant \varepsilon \leqslant 1)$, and for $H_{uu}(\bar{x}, \bar{u}, \lambda; t)$ positive-definite, a reduction in cost at each iteration is predicted [1].

A.4. DIFFERENCES BETWEEN THE ALGORITHMS OF SECTIONS A.2 AND A.3

There are some notable differences between the algorithms:

1. The successive-sweep algorithm requires the integration of an extra set of differential equations; i.e., the $n-\dot{h}$ equations.
2. The variable $\lambda(t)$ is, in general, not equal to $V_x(t)$ because of the different forms of the $\dot{\lambda}$ and $\dot{V}_x$ equations; this implies that

$$H(\bar{x}, \bar{u}, \lambda; t) \neq H(\bar{x}, \bar{u}, V_x; t) \tag{A.4.1}$$

3. The matrix Riccati equation for $\dot{P}$, though of the same form as that for $\dot{V}_{xx}$, produces a $P(t) \neq V_{xx}(t)$ for the reason given in 2.
4. The expression for u, Equation (A.2.7), is not equivalent to Equation (A.3.6) because of 2 and 3, above. However, we note that on an optimal trajectory:
5. $H_u(t) = 0 \Rightarrow h(t) = 0$; $\quad t \in [t_o, t_f]$
6. $H_u(t) = 0 \Rightarrow \lambda(t) = V_x(t)$; $\quad t \in [t_o, t_f]$
7. Point 6 $\Rightarrow P(t) = V_{xx}(t)$; $\quad t \in [t_o, t_f]$
8. The expressions for u are identical. That is, as the optimal trajectory is approached (i.e., as $H_u \to 0$), the methods become identical. Further, we note that, for the LQP problem:
9. H_{xx}, H_{ux}, and H_{uu} are not functions of V_x or λ; so the Riccati equations are independent of V_x and λ; i.e., $P(t) = V_{xx}(t)$; $t \in [t_o, t_f]$.
10. The $\dot{h}$ equation adds directly to the $\dot{\lambda}$ equation to yield $\dot{V}_x$. This addition cannot be performed for a nonlinear system because H_{uu}^{-1} and H_{ux} are functions of V_x.

11. In view of 9 and 10 the expressions for u are equivalent. The two algorithms are thus identical for the LQP problem.

From the above points, it is clear that the two methods are not identical when applied to nonlinear control problems, if the nominal control is nonoptimal.

In [1] it is shown that the successive-sweep method is actually an approximation to Mayne's algorithm. Computationally, it has been shown that Mayne's algorithm converges faster than the succesive-sweep method on a nonlinear control problem [1].

A.5. CONCLUSION

We have indicated some differences between a differential dynamic programming algorithm due to Mayne, and the successive-sweep method. It has been shown [1] that not only are the algorithms different in form, but the successive-sweep method is an approximation to Mayne's algorithm. We thus have the interesting situation that an approximate method actually requires the integration of more differential equations than the accurate differential dynamic programming algorithm.

The methods described above have the following requirements:

1. $H_{uu}^{-1}(\bar{x}, \bar{u}, V_x; t)$ and $H_{uu}^{-1}(\bar{x}, \bar{u}, \lambda; t)$ must be positive-definite for $t \in [t_0, t_f]$

2. Inequality constraints on control variables cannot be handled directly; they have to be approximated by penalty functions.

3. The first requirement excludes the bang-bang type of control problem, where $H_{uu} = 0$.

The strong variation algorithms presented in Chapters 2 and 3 do not suffer from these drawbacks.

References

1. D. H. Jacobson, *Intern. J. Control*, **7**, 175 (1968).
2. D. Q. Mayne, *Intern. J. Control*, **3**, 85 (1966).
3. S. K. Mitter, *Automatica*, **3**, 135 (1966).
4. S. R. McReynolds and A. E. Bryson, *Proc. 6th Joint Auto. Control Conf., Troy, New York, 1965*, p. 551.
5. S. R. McReynolds, *J. Math. Anal. Appl.*, **19**, 565 (1967).

Appendix B

ERROR ANALYSIS FOR BANG-BANG ALGORITHMS

B.1. INTRODUCTION

In Chapter 3, new algorithms for determining optimal bang-bang control are described. Here, by means of an error analysis [1], we provide justification for the neglect of the $V_{xxx}\delta x \delta x$ term in Equation (3.3.1). The error analysis is somewhat different from that given in Section 2.2.1 for non-bang-bang problems.

B.2. ERROR ANALYSIS

If the term $V_{xxx}\delta x \delta x$ is included in the Bellman equation expansion, the following differential equations are obtained:

$$\begin{aligned} -\dot{a} &= H - H(\bar{x}, \bar{u}, V_x; t) \\ -\dot{V}_x &= H_x + V_{xx}(f - f(\bar{x}, \bar{u}\ t)) \\ -\dot{V}_{xx} &= H_{xx} + f_x{}^T V_{xx} + V_{xx} f_x + \tfrac{1}{2} V_{xxx}(f - f(\bar{x}, \bar{u}; t)) \\ &\quad + \tfrac{1}{2}(f - f(\bar{x}, \bar{u}; t))^T V_{xxx} \end{aligned} \tag{B.2.1}$$

Unless specified otherwise all quantities in Equations (B.2.1) are evaluated at $\bar{x}, u^*; t$, where

$$u^* = \arg\min_u H(\bar{x}, u, V_x; t) \tag{B.2.2}$$

If the V_{xxx} terms in Equation (B.2.1) are neglected, then $\Delta a(t)$—the error in the predicted change in cost—is clearly of order

$$\int_{t_f}^{t} \int_{t_f}^{t_3} \int_{t_f}^{t_2} |u^*(t_1) - \bar{u}(t_1)|\, dt_1\, |u^*(t_2) - \bar{u}(t_2)|\, dt_2\, |u^*(t_3) - \bar{u}(t_3)|\, dt_3 \tag{B.2.3}$$

while $a(t)$ is of order

$$\int_{t_f}^{t} |u^*(t_4) - \bar{u}(t_4)|\, dt_4 \tag{B.2.4}$$

if the problem is nonsingular. For $|u^* - \bar{u}|$ of order ε, $\Delta a(t)$ is of order ε^3, and $a(t)$ is of order ε. Alternatively, for $(t_f - t)$ of order ε, $\Delta a(t)$ is of order ε^3, and $a(t)$ is of order ε. In either case $a(t)$ is an estimate of the true predicted change in cost, the error in $a(t)$ being third-order. Owing to the neglect of the V_{xxx} terms, an error in the switch times of u^*, of magnitude

$$\Delta t_s = -V_{t_s t_s}^{-1} f_u^T \Delta V_x|_{t_s} \tag{B.2.5}$$

is introduced. However, ΔV_x—the error in V_x—is clearly second-order in ε. The error introduced into $a(t_s)$ is, to second-order in Δt_s (see Chapter 3):

$$V_{t_s} \Delta t_s + \tfrac{1}{2} V_{t_s t_s} \Delta t_s^2 \tag{B.2.6}$$

At a switch time, $V_{t_s} = 0$ (see Chapter 3); thus the error introduced into $a(t_s)$ by Δt_s is of order ε^4. The upshot of this argument is that the neglect of the V_{xxx} terms in the V_{xx} equation introduces a third-order error into the estimate of the predicted reduction in cost. The step-size adjustment method described in Chapter 2 ensures that this third-order error term is negligible; and hence that a reduction in cost occurs, by applying the new control function, at each iteration, only over a time interval $[t_1, t_f]$, where t_1 is chosen to be sufficiently close to t_f.

Reference

1. D. H. Jacobson, *IEEE Trans. Auto. Control*, **AC-14**, 197 (1969).

AUTHOR INDEX

Numbers in parentheses indicate the numbers of the references when these are cited in the text without the name of the author.

Numbers set in *italics* designate those page numbers on which the complete literature citations are given.

SUBJECT INDEX